SiGe Heterojunction Bipolar Transistors

SiGe Heterojunction Bipolar Transistors

Peter Ashburn
University of Southampton, Southampton, UK

John Wiley & Sons, Ltd

Telephone (+44) 1243 779777

Email (for orders and customer service enquiries): cs-books@wiley.co.uk
Visit our Home Page on www.wileyeurope.com or www.wiley.com

Other Wiley Editorial Offices

John Wiley & Sons Inc., 111 River Street, Hoboken, NJ 07030, USA

Jossey-Bass, 989 Market Street, San Francisco, CA 94103-1741, USA

Wiley-VCH Verlag GmbH, Boschstr. 12, D-69469 Weinheim, Germany

John Wiley & Sons Australia Ltd, 33 Park Road, Milton, Queensland 4064, Australia

John Wiley & Sons (Asia) Pte Ltd, 2 Clementi Loop #02-01, Jin Xing Distripark, Singapore 129809

John Wiley & Sons Canada Ltd, 22 Worcester Road, Etobicoke, Ontario, Canada M9W 1L1

Wiley also publishes its books in a variety of electronic formats. Some content that appears in print may not be available in electronic books.

Library of Congress Cataloging-in-Publication Data

Ashburn, Peter.
SiGe heterojunction bipolar transistors / Peter Ashburn.
p. cm.
Includes bibliographical references and index.
ISBN 0-470-84838-3
1. Bipolar transistors. 2. Silicon. 3. Germanium. I. Title.

TK7871.96.B55A88 2003
621.3815′282 – dc22
2003049482

British Library Cataloguing in Publication Data

A catalogue record for this book is available from the British Library

ISBN 0-470-84838-3

Typeset in 10.5/13pt Sabon by Laserwords Private Limited, Chennai, India
Printed and bound in Great Britain by TJ International, Padstow, Cornwall
This book is printed on acid-free paper responsibly manufactured from sustainable forestry in which at least two trees are planted for each one used for paper production.

To my wife, Ann, and daughters Jenny and Susie

Contents

Preface

In the late 1980s silicon bipolar technologies were reaching maturity, with values of cut-off frequency f_T around 30 GHz and ECL gate delays between 20 and 30 ps. The 1990s saw remarkable developments as the silicon-germanium heterojunction bipolar transistor (HBT) emerged from research labs around the world and entered production in mainstream radio frequency BiCMOS technologies. These developments have had a dramatic impact on the performance on bipolar transistors and have led to values of f_T approaching 400 GHz and ECL gate delays below 5 ps. SiGe BiCMOS technology is seriously challenging III/V and II/VI technologies in high-frequency electronics applications, such as mobile communications and optical fibre communications. Furthermore, the success of silicon-germanium in bipolar technologies has paved the way for the use of silicon-germanium in CMOS technologies. A similar revolution is now underway in the design of MOS transistors as silicon-germanium is used to give improved channel mobility in a number of different types of heterojunction MOSFET.

The purpose of this book is to bring together in a single text all aspects of the physics and technology of silicon bipolar transistors and silicon-germanium heterojunction bipolar transistors. The book covers the basic DC and AC transistor operation, as well as important second-order effects that influence transistor performance. A number of relevant materials topics are covered, including the diffusion of boron and arsenic in silicon, the properties of silicon-germanium and polysilicon, strain effects in silicon-germanium, and the epitaxial growth of silicon and silicon-germanium. The fabrication of silicon bipolar transistors and SiGe HBTs is covered in detail and self-aligned schemes for the fabrication of both types of device are presented. Accurate circuit simulation is crucially important to the successful design of bipolar and BiCMOS circuits, and hence compact models of bipolar transistors are explained in detail and related to the physical transistor operation. The book concludes with coverage of overall bipolar technology optimization, which allows the transistor design, technology specification and circuit design to be optimized to give minimum ECL and CML gate delay. The book is intended primarily for practising engineers and scientists and for students at the masters and postgraduate level.

In the first chapter the reader is given an overview of silicon and SiGe heterojunction bipolar technologies and is introduced to the operating principles of the bipolar transistor. A more rigorous and quantitative description of the DC bipolar transistor operation is then given in the succeeding two chapters. Chapter 2 deals with the basic physics of the bipolar transistor and takes the reader through the derivation of an expression for the current gain. Heavy doping effects have a strong effect on the current gain and are covered in detail in Chapter 3. Chapter 4 describes second-order effects that influence bipolar transistor operation at the extremes of currents and voltages. The high-frequency performance of the bipolar transistor is described in Chapter 5, including descriptions of the cut-off frequency f_T and the maximum oscillation frequency $f_{\max}$, and physical explanations of the Kirk effect, quasi-saturation and current crowding.

Chapters 6, 7 and 8 deal with more recent developments that have had a strong impact on bipolar transistor performance. Chapter 6 covers polysilicon emitters from both the technological and device physics points of view. A simple expression for the base current of a polysilicon emitter is derived and the practical design of polysilicon emitters is covered in detail. Chapter 7 summarizes the materials and physical properties of silicon-germanium and the epitaxial growth of both silicon and silicon-germanium. Silicon-germanium HBTs are discussed in Chapter 8 and it is shown that the device operation can be understood using simple developments of the theory in Chapters 2 to 5. The performance of SiGe HBTs is limited by the diffusion of boron in the base and so the mechanisms involved in boron diffusion are described. The use of carbon doping in the silicon-germanium to reduce boron diffusion is explained.

Chapters 9 and 10 deal with silicon bipolar and silicon-germanium heterojunction bipolar technologies. The key processing steps required to fabricate a bipolar transistor are identified and discussed in detail in Chapter 9. These include buried layer, epitaxy, isolation, selective-implanted-collector, base and emitter. Examples are then given of four types of bipolar process: double polysilicon self-aligned bipolar, single polysilicon bipolar, complementary bipolar and BiCMOS. Silicon-germanium heterojunction bipolar technology is introduced in Chapter 10 and the two approaches of differential epitaxy and selective epitaxy are outlined. Silicon-germanium-carbon HBT processes and germanium implanted HBT processes are also described. The main application of SiGe HBT technologies is in radio frequency circuits and so integrated circuit passives are described, including resistors, capacitors, inductors, and varactor diodes.

Chapters 11 and 12 describe the use of bipolar transistors and SiGe HBTs in circuits. Chapter 11 describes compact bipolar transistor models, beginning with the Ebers-Moll model and building towards the Gummel-Poon model in easy-to-understand stages. The well known SPICE2G bipolar transistor model is described in detail and the chapter concludes with consideration of the VBIC95 and Mextram bipolar transistor models. In Chapter 12 optimization of the overall process, transistor and circuit design is discussed using a quasi-analytical expression for the gate delay of an ECL logic gate in terms of all the time constants of the circuit. The application of the gate delay expression is demonstrated by case studies for the double polysilicon self-aligned bipolar technology and the SiGe HBT technology.

Many people have contributed directly and indirectly to the writing of this book, and it would be impossible to find the space to thank them all. Nevertheless, I would like to identify a number of colleagues who have made particularly large contributions to this project. First, acknowledgements should go to my colleagues in the Microelectronics Group at Southampton University, with whom I have had numerous stimulating discussions about device physics. These include Henri Kemhadjian, Greg Parker, Arthur Brunnschweiler, Alan Evans, Kees de Groot and Darren Bagnall. A debt of gratitude is also owed to my past and present research students, who have contributed greatly to my understanding of device physics in general and bipolar transistors in particular. These include Bus Soerowirdjo, Alan Cuthbertson, Eng Fong Chor, Graham Wolstenholme, Nasser Siabi-Shahrivar, Ian Post, Alan Shafi, Wen Fang, Nick Moiseiwitsch, Jochen Schiz, Iain Anteney, Michele Mitchell, Huda El Mubarek, Dominik Kunz and Enrico Gili. Particular thanks are due to Kees de Groot for checking the first draft of my book.

Finally, no list of acknowledgements would be complete without mention of my wife and family for their support during the execution of this seemingly endless task. I will therefore finish by acknowledging the patience and support of my wife Ann, and children Jennifer and Susan.

Peter Ashburn
Southampton, England
April 2003

Physical Constants and Properties of Silicon and Silicon-Germanium

PHYSICAL CONSTANTS

Quantity	Value
Boltzmann's constant (k)	1.38×10^{-23} JK^{-1}
Electronic charge (q)	1.602×10^{-19} C
Permittivity of free space (ε_0)	8.85×10^{-12} C^2/Nm
Planck's constant (h)	6.626×10^{-34} Js
Free electron mass (m_o)	9.108×10^{-31} kg
Electron-volt (eV)	1.602×10^{-19} J

PROPERTIES OF SILICON AND SILICON-GERMANIUM

Value	Silicon	Silicon-germanium
Lattice constant (nm)	0.543	$a_{SiGe} = 0.543 + x(0.566 - 0.543)$
Bandgap (eV)	1.170	$E_G(x) = 1.17 - 0.96x + 0.43x^2 - 0.17x^3$
Dielectric constant	11.9	$\varepsilon(x) = 11.9(1 + 0.35x)$
Density N_C of states in the conduction band at 300 K (cm^{-3})	2.8×10^{19}	2.8×10^{19}

Value	Silicon	Silicon-germanium
Density N_V of states in the valence band at 300 K (cm^{-3})	1.04×10^{19}	Figure 7.8
Apparent bandgap narrowing in the base	Figure 3.7	Figure 7.9
Apparent bandgap narrowing in the emitter	Figure 3.6	–
Critical thickness	–	Figure 7.3

List of Symbols

a	Lattice constant
A	Area of the emitter/base junction
A_e	Modified Richardson constant
α	common base current gain
α_R	Reverse common base current gain
α_F	Forward common base current gain
α_T	Base transport factor
b_b	Width of the extrinsic base region of a bipolar transistor
b_c	Width of the buried layer of a bipolar transistor
b_e	Width of the emitter of a bipolar transistor
BV	Breakdown voltage
BV_{CBO}	Bipolar transistor breakdown voltage between the collector and base with the emitter open-circuit
BV_{CEO}	Bipolar transistor breakdown voltage between the collector and emitter with the base open-circuit
B_s^-	Substitutional boron atom
B_i^-	Negatively charged boron interstitial pair
B_i^0	Neutral boron interstitial pair
β	Common emitter current gain
β_F	Forward common emitter current gain
β_R	Reverse common emitter current gain
C_{DC}	Collector diffusion capacitance
C_{DE}	Emitter diffusion capacitance
C_{JE}	Emitter/base depletion capacitance
C_{JC}	Base/collector depletion capacitance

C_{JCI}	Intrinsic collector/base depletion capacitance
C_{JCX}	Extrinsic collector/base depletion capacitance
C_{JS}	Collector/substrate depletion capacitance
C_{μ}	Collector/base capacitance in the small-signal hybrid-π model
C_{π}	Emitter/base capacitance in the small-signal hybrid-π model
C_N	Auger recombination coefficient
C_L	Load capacitance due to interconnections
C_S	Concentration of reactant gas at the surface of the film
C_G	Concentration of reactant gas in the bulk of the gas
C_T	Total number of reactant molecules per unit volume of gas
C_s	Substitutional carbon atom
C_i	Interstitial carbon atom
χ_e	Effective barrier height for electron tunnelling
χ_h	Effective barrier height for hole tunnelling
D_i	Intrinsic diffusion coefficient for dopant diffusion with a neutral point defect
D^-	Intrinsic diffusion coefficient for dopant diffusion with a singly charged acceptor point defect
D^+	Intrinsic diffusion coefficient for dopant diffusion with a singly charged donor point defect
D_B	Diffusion coefficient of boron
D_n	Diffusion coefficient of electrons
D_p	Diffusion coefficient of holes
D_{nb}	Diffusion coefficient of electrons in the base
D_{pe}	Diffusion coefficient of holes in the emitter
D_G	Diffusion coefficient of the reactant species in a gas
ΔE_c	Conduction band discontinuity in a heterojunction
ΔE_v	Valence band discontinuity in a heterojunction
ΔE_{gb}	Apparent bandgap narrowing in the base
ΔE_{ge}	Apparent bandgap narrowing in the emitter
ΔE_G	Bandgap narrowing due to germanium in the base
ΔV	Logic swing of an ECL or CML gate
δ	Interfacial layer thickness in a polysilicon emitter
E	Electric field
E_{crit}	Critical electric field for avalanche breakdown
E_F	Fermi level
E_{Fn}	Electron quasi-Fermi level
E_{Fp}	Hole quasi-Fermi level
E_C	Energy level of the conduction band

E_V	Energy level of the valence band
E_G	Semiconductor bandgap
E_i	Intrinsic fermi level
E_t	Energy level of a deep level in the bandgap
E_B	Activation energy for boron diffusion
e_n	Emission probability for electrons at a deep level
e_p	Emission probability for holes at a deep level
ε_0	Permittivity of free space
ε_r	Relative permittivity or dielectric constant of silicon
F	Friction
F_1	Flux of reactant species
f_T	Cut-off frequency
f_{TMAX}	Peak value of the cut-off frequency
$f_{\max}$	Maximum oscillation frequency
G_b	Base Gummel number
G_e	Emitter Gummel number
G_n	Electron generation rate
G_p	Hole generation rate
G_R	Growth rate
g_m	Transconductance
γ	Emitter efficiency
γ_M	Mole fraction of reactant species
h	Planck's constant
h_{FE}	Common emitter current gain
h_G	Gas phase mass transport coefficient
I	Interstitial
I_B	Base current
I_C	Collector current
I_E	Emitter current
I_S	Saturation current
I_{ES}	Emitter saturation current
I_{CS}	Collector saturation current
I_{pe}	Hole diffusion current in the emitter
I_{ne}	Electron diffusion current at the emitter edge of the base
I_{nc}	Electron diffusion current at the collector edge of the base
I_{rb}	Recombination current in the base
I_{rg}	Recombination current in the emitter/base depletion region
I_{gen}	Generation current in a reverse biased depletion region

J_n	Electron current density
J_p	Hole current density
k	Boltzmann's constant
k_S	Surface reaction rate constant
L_n	Electron diffusion length
L_p	Hole diffusion length
L_{nb}	Electron diffusion length in the base
L_{pe}	Hole diffusion length in the emitter
l_b	Length of the extrinsic base region of a bipolar transistor
l_c	Length of the buried layer of a bipolar transistor
l_e	Length of the emitter of a bipolar transistor
M	Avalanche breakdown multiplication factor
m	Base current ideality factor
m_e^*	Electron effective mass
m_h^*	Hole effective mass
μ_n	Electron mobility
μ_p	Hole mobility
N_a	Acceptor concentration
N_d	Donor concentration
N_{ab}	Acceptor concentration in the base
N_{dc}	Donor concentration in the collector
N_{de}	Donor concentration in the emitter
N_{deff}	Effective doping concentration, including the effects of bandgap narrowing
N_C	Effective density of states in the conduction band
N_V	Effective density of states in the valence band
N_t	Density of deep levels
N_F	Number of atoms incorporated into a unit volume of a growing film
n	Electron concentration
n_b	Electron concentration in the base
n_{bo}	Equilibrium electron concentration in the base
n_i	Intrinsic carrier concentration
n_{io}	Intrinsic carrier concentration in a lightly doped semiconductor
n_{ie}	Intrinsic carrier concentration in a heavily doped emitter
n_{ib}	Intrinsic carrier concentration in a heavily doped base

p	Hole concentration
p_e	Hole concentration in the emitter
p_{eo}	Equilibrium hole concentration in the emitter
Q	Stored charge
Q_b	Charge stored in the base
Q_e	Charge stored in the emitter
q	Charge on an electron
R_e	Reynolds number
R_B	Base resistance
R_{BI}	Intrinsic base resistance
R_{BX}	Extrinsic base resistance
R_C	Collector resistance
R_E	Emitter resistance
R_{EF}	Emitter follower resistor in an ECL circuit
R_L	Load resistor in an ECL or CML circuit
R_{SBI}	Sheet resistance of the intrinsic base
R_{SBX}	Sheet resistance of the extrinsic base
R_{SBL}	Sheet resistance of the buried layer
R_{CON}	Contact resistance
ρ_G	Density of a gas
S_M	Surface recombination velocity of a metal contact
S_P	Effective recombination velocity at the edge of the polysilicon layer in a polysilicon emitter
S_{EFF}	Effective recombination velocity for a complete polysilicon emitter
S_I	Effective recombination velocity due to recombination at traps at the polysilicon/silicon interface
σ_n	Capture cross-section for electrons
σ_p	Capture cross-section for holes
T	Temperature
τ_n	Electron lifetime
τ_p	Hole lifetime
τ_{nb}	Electron lifetime in the base
τ_{pe}	Hole lifetime in the emitter
τ_A	Auger lifetime
τ_F	Forward transit time
τ_R	Reverse transit time

τ_E Emitter delay
τ_{EBD} Emitter/base depletion region delay
τ_B Base transit time
τ_{CBD} Collector/base depletion region transit time
τ_{RE} Delay due to the emitter/base and collector/base depletion capacitances

τ_D Propagation delay
U Recombination rate
U_n Electron recombination rate
U_p Hole recombination rate

V Vacancy
V_{BE} Base/emitter voltage
V_{BC} Base/collector voltage
V_{CE} Collector/emitter voltage
V_{AF} Forward Early voltage
V_{AR} Reverse Early voltage
V_{bi} Built-in voltage of a *p-n* junction
V_{JE} Built-in voltage of E/B junction
v_{th} Thermal velocity
v_{scl} Scattering limited velocity
v_{isc} Viscosity of a gas

W_B Basewidth
W_E Depth of the emitter
W_D Depletion width
W_{CBD} Collector/base depletion width

1

Introduction

1.1 EVOLUTION OF SILICON BIPOLAR TECHNOLOGY

The bipolar transistor was invented by a team of researchers at the Bell Laboratories, USA, in 1948 [1]. The original transistor was a germanium point contact device, but in 1949 Shockley published a paper on *pn* junctions and junction transistors [2]. These two papers laid the foundations for the modern bipolar transistor, and made possible today's multi-million dollar microelectronics industry.

A large number of innovations and breakthroughs were required to convert the original concept into a practical technology for fabricating VLSI circuits. Among these, diffusion was an important first step, since it allowed thin bases and emitters to be fabricated by diffusing impurities from the vapour phase [3]. The use of epitaxy [4] to produce a thin single-crystal layer on top of a heavily doped buried layer was also a big step forward, and led to a substantial reduction in the collector series resistance. Faster switching speeds and improved high-frequency gain were the main consequences of this innovation.

The next stage in the evolution of bipolar technology was the development of the planar process [5], which allowed bipolar transistors and other components, such as resistors, to be fabricated simultaneously. This is clearly necessary if circuits are to be produced on a single silicon chip (i.e. integrated circuits). Figure 1.1 shows the main features of a basic planar bipolar process. Electrical isolation between adjacent components is provided by a *p*-type isolation region, which is diffused

SiGe Heterojunction Bipolar Transistors Peter Ashburn
 ISBN: 0-470-84838-3

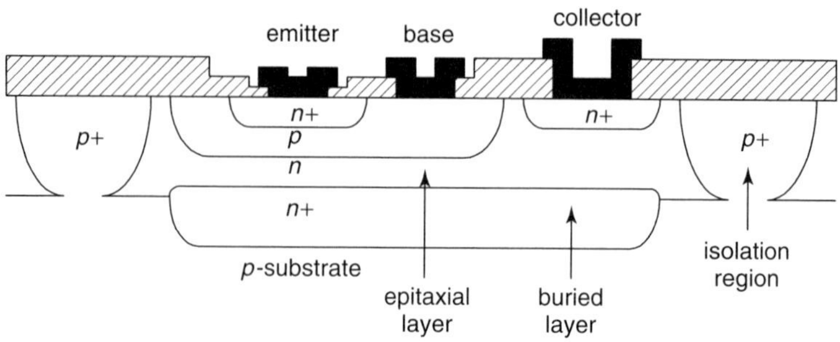

Figure 1.1 Cross-sectional view of a basic, planar, integrated circuit, bipolar transistor

from the surface to intersect the *p*-substrate. For the isolation to be effective, the diffusion must completely surround the device, and the isolation junction must be reverse biased by connecting the *p*-substrate to the most negative voltage in the circuit. The *n*+ diffusion underneath the collector contact is needed to give a low-resistance ohmic contact. This type of transistor typically had a cut-off frequency f_T of around 500 MHz, and was used to produce the early TTL circuits and operational amplifiers.

In the 1970s and 1980s major innovations in silicon technology were introduced that led to considerable improvements in bipolar transistor performance. Ion implantation was used to improve the uniformity and reproducibility of the base [6] and emitter [7] regions, and also to produce devices with narrower basewidths [8]. Furthermore, the use of polysilicon emitters [9] and self-aligned processing techniques [10] revolutionized the design of silicon bipolar transistors and led to the development of the self-aligned double polysilicon bipolar transistor.

Figure 1.2 shows a cross-section of a typical double polysilicon bipolar transistor. It can be seen that it bears little resemblance to the more traditional transistor in Figure 1.1. Contact to the emitter is made via an *n*+ polysilicon emitter and to the base via a *p*+ polysilicon layer. The emitter and extrinsic base regions are separated by an oxide spacer on the sidewall of the *p*+ polysilicon, which allows the emitter to be self-aligned to the extrinsic base. The junction isolation of Figure 1.1 has been replaced by a combination of oxide isolation and deep trench isolation. The base region is butted against the oxide isolation region, and hence gives a much lower parasitic collector/base capacitance. An *n*+ collector sink is used to contact the buried layer to further reduce the collector resistance. The double polysilicon bipolar transistor is a high-frequency bipolar transistor with a cut-off frequency f_T of around 30 GHz, and is typically used in emitter coupled logic circuits and high-frequency analogue

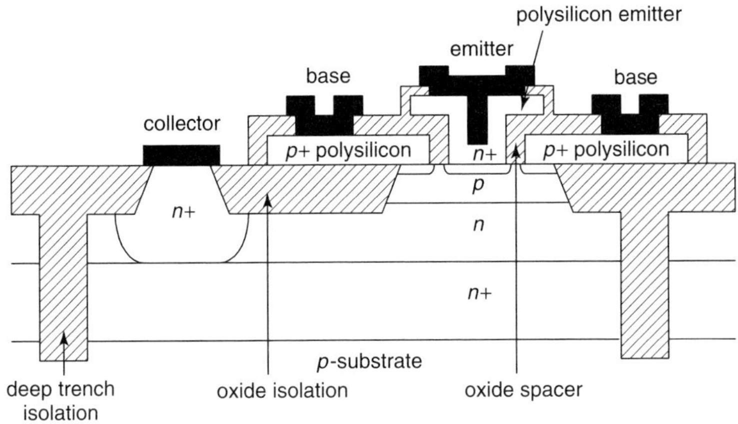

Figure 1.2 Cross-sectional view of a self-aligned double polysilicon bipolar process

circuits. ECL gate delays approaching 10 ps [11] have been achieved in circuits incorporating double polysilicon bipolar transistors.

For many applications, there are many benefits to be obtained by combining bipolar and MOS transistors on a single chip [12]. The main motivation in digital circuits for moving from CMOS to BiCMOS technology is that bipolar transistors can sink a larger current per unit device area than MOS transistors. They are therefore more effective in driving the large on-chip capacitances that are commonly encountered in digital VLSI systems [13]. BiCMOS processes also allow high-speed digital circuits to be combined on the same chip as high-performance analogue circuits [14], thereby producing a technology capable of integrating a wide variety of mixed signal systems.

1.2 EVOLUTION OF SILICON-GERMANIUM HBT TECHNOLOGY

In the 1990s a further revolution in bipolar transistor design occurred with the emergence of SiGe Heterojunction Bipolar Transistors (HBTs). Previously, heterojunction bipolar transistors had only been available in compound semiconductor technologies, such as AlGaAs/GaAs [15], because effective heterojunction formation requires two semiconductors with similar lattice spacing, as is the situation for AlGaAs and GaAs. The lattice mismatch between Si and Ge is relatively large at 4.2%, and hence it is very difficult to form a heterojunction between Si and SiGe without the generation of misfit dislocations at the interface.

However, materials research carried out in the 1980s showed that a good heterojunction could be obtained if the SiGe layer was thin and the Ge content relatively low (below 30%). In these circumstances, the SiGe layer grows under strain so that it fits perfectly onto the silicon lattice without the generation of misfit dislocations. The epitaxial growth of reproducible strained, or pseudomorphic, SiGe layers was the vital technology breakthrough that led to the emergence of the SiGe HBT [16–19].

Figure 1.3 shows a cross-section of a typical SiGe heterojunction bipolar transistor [20]. The *p*+ SiGe base layer is grown after oxide isolation formation and is followed in the same growth step by the growth of a *p*-type Si cap. Single-crystal material is formed where the silicon collector is exposed and polycrystalline material over the oxide isolation. The boundary between these two types of material is shown by the dotted lines in Figure 1.3. The polycrystalline material is heavily *p*+ doped using an extrinsic base implant and then used to contact the base in a similar way to that employed in the double polysilicon bipolar transistor in Figure 1.2. The emitter is formed by diffusing arsenic from the polysilicon emitter to over-dope the Si cap *n*-type.

SiGe HBTs have been produced with values of f_T and f_{max} of over 300 GHz, and with extremely low values of noise figure. Their main applications are in wireless communication systems and optical fibre communication systems. SiGe HBTs are generally integrated with MOS transistors in a BiCMOS technology, so that the HBTs are used in the RF

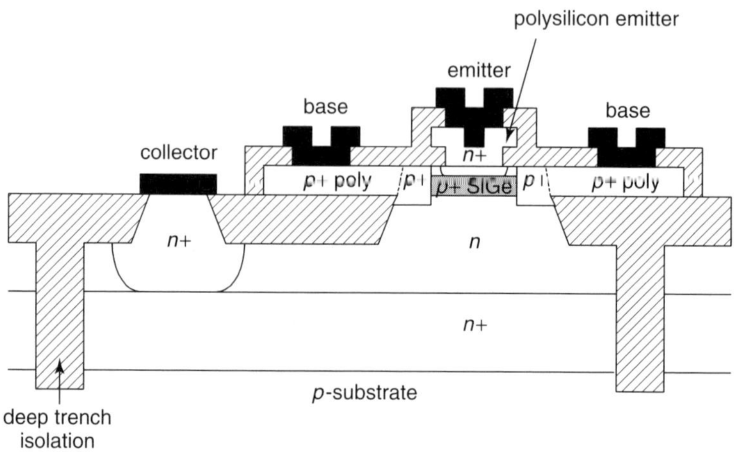

Figure 1.3 Cross-sectional view of a silicon-germanium heterojunction bipolar transistor

circuits and the MOS transistors in the digital CMOS circuits. BiCMOS technologies incorporating SiGe HBTs are therefore ideally suited for producing RF systems on a single chip.

1.3 OPERATING PRINCIPLES OF THE BIPOLAR TRANSISTOR

For the purposes of understanding the operation of the bipolar transistor, the structures in Figures 1.1–1.3 can be considered as essentially one-dimensional, as illustrated in Figure 1.4(a). Although this is clearly an approximation, it is valid over a remarkable range of operating conditions. In practice, it only begins to break down at very high current levels when the series resistances of the transistor and high current effects become important. In the first instance, an idealized bipolar transistor will be assumed in which the doping profiles are uniform, as illustrated in Figure 1.4(b). In practice, this is a good approximation for Si/SiGe heterojunction bipolar transistors, but in silicon transistors the profiles are generally Gaussian. The implications of this deviation from ideality will be considered in Section 3.6.

The band diagram for our idealized, silicon bipolar transistor is shown in Figure 1.4(c). In the absence of any applied bias, the Fermi level E_F is constant throughout the device. The Fermi level E_F and the intrinsic Fermi level E_i are related to the carrier concentrations in the emitter, base and collector through the following equations [21]:

$$n = N_C \exp\left[-\frac{E_C - E_F}{kT}\right] = n_i \exp\left[\frac{E_F - E_i}{kT}\right] \tag{1.1}$$

$$p = N_V \exp\left[-\frac{E_F - E_V}{kT}\right] = n_i \exp\left[\frac{E_i - E_F}{kT}\right] \tag{1.2}$$

where N_C and N_V are the conduction band and valence band density of states. From equations (1.1) and (1.2), it can be seen that the product of the electron and hole concentration in a given region of the transistor is a constant, given by:

$$pn = n_i^2 \tag{1.3}$$

The relationship between the doping profiles in Figure 1.4(b) and the band diagram in Figure 1.4(c) is now clear. In particular, the electron and hole concentrations given in equations (1.1) and (1.2) are directly

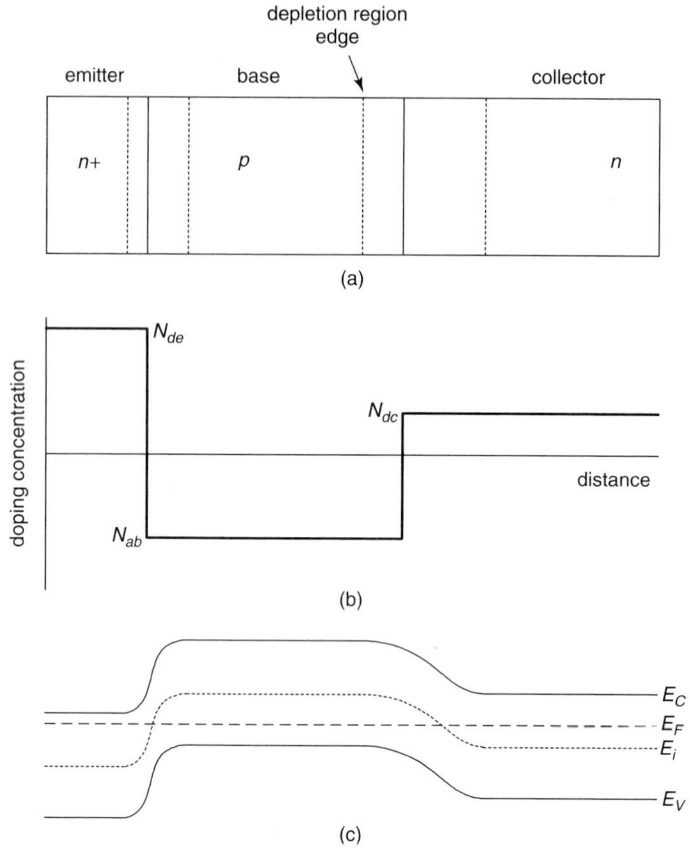

Figure 1.4 (a) One-dimensional representation of an *npn* bipolar transistor; (b) doping profiles for the case of abrupt *pn* junctions; (c) band diagram for a transistor with no applied bias

related to the separation between the Fermi level and the intrinsic Fermi level, as drawn in Figure 1.4(c).

In order to use a bipolar transistor in practical circuits, external bias must be applied to the emitter/base and collector/base junctions. These two junctions provide four possible bias configurations, as illustrated in Figure 1.5. The forward active mode of operation is the most useful, because in this configuration the gain of the transistor can be exploited to produce current amplification. A forward bias of approximately 0.6 V is applied to the base/emitter junction and a reverse bias to the collector/base junction.

The band diagram for the forward active region of operation is shown in Figure 1.6. The applied bias leads to a separation of the Fermi levels

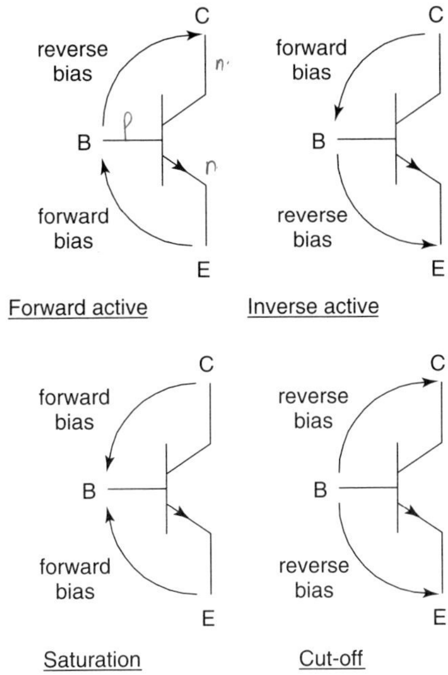

Figure 1.5 The four regions of operation of a bipolar transistor

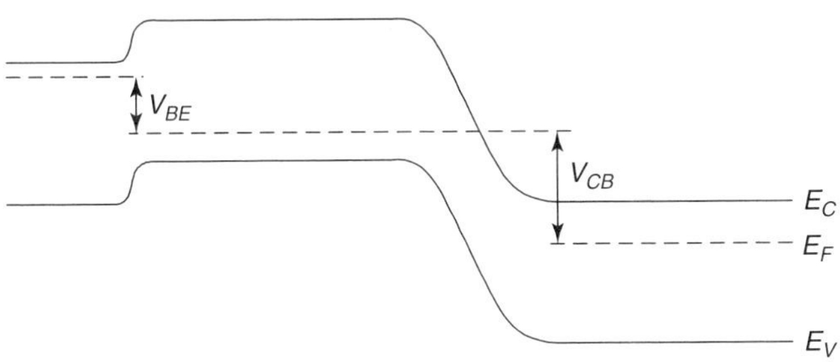

Figure 1.6 Band diagram for a bipolar transistor biased in the forward active region

in the emitter, base and collector, with the separation being equal to the applied bias. The forward bias across the emitter/base junction leads to a decrease in the potential barrier between the emitter and base, whereas the reverse bias across the collector/base junction leads to an increase in the potential barrier between the base and collector.

The other bias configurations in Figure 1.5 are also often encountered in practical circuits. In the inverse (or reverse) active mode, the emitter/base junction is reverse biased and the collector/base forward biased. This arrangement is less useful than the forward active mode because the inverse gain of the transistor is very low, though it is used in I^2L circuits [22]. In the cut-off mode both junctions are reverse biased, and hence no current can flow between emitter and collector. The transistor is therefore off, and behaves like an open switch. Conversely, in the saturation mode both junctions are forward biased, which enables a large current to flow between emitter and collector. In this configuration the transistor can be viewed as a closed switch.

The electrical properties of a bipolar transistor can be characterized by a number of electrical parameters, the most important of which is the common emitter current gain β. This is the ratio of collector current to base current, and is given by:

$$\beta = \frac{I_C}{I_B} \tag{1.4}$$

In a typical commercial transistor the collector current is approximately one hundred times larger than the base current, giving a current gain of around 100. In order to understand how this important property of the bipolar transistor arises we must consider how it functions when external bias is applied.

In the forward active mode, the forward biasing of the emitter/base junction causes a large number of electrons to be injected from the emitter into the base. A concentration gradient is therefore established in the base, which encourages the electrons to diffuse towards the collector. If the base of the transistor were very wide all the injected electrons would recombine before reaching the collector, and the transistor would merely behave like two back-to-back diodes. However, the essence of the bipolar transistor is that the base is sufficiently narrow that the majority of electrons reach the collector/base junction, where they are swept across into the collector by the large electric field across the reverse biased junction. This is achieved by making the basewidth comparable with, or smaller than, the diffusion length of electrons in the base. The base current is determined by the number of holes injected from the base into the emitter. The base current can be made much smaller than the emitter current by doping the emitter much more heavily than the base.

A related electrical parameter to the common emitter current gain is the common base current gain α, which is the ratio of the collector current to the emitter current:

$$\alpha = \frac{I_C}{I_E} \tag{1.5}$$

The emitter current is given by the sum of the collector and base currents:

$$I_E = I_B + I_C \tag{1.6}$$

It is therefore apparent that α and β are related by:

$$\alpha = \frac{\beta}{1+\beta} \tag{1.7}$$

The common emitter and common base current gains can be measured by biasing the transistor into the forward active region and taking readings of base, emitter and collector current. Three alternative circuit configurations are possible, depending upon which terminal is common between the input and output. These are illustrated in Figure 1.7, and are termed the common emitter, common base and common collector circuit configurations.

The common emitter current gain β is obtained by connecting the transistor in the common emitter configuration, as illustrated in Figure 1.7(a), and plotting the collector current as a function of collector/emitter voltage, with the base current as a parameter. The resulting characteristic is illustrated in Figure 1.8. The common emitter current gain is obtained by reading off the value of collector current obtained for one of the values of base current and taking the ratio.

The common base current gain α can be measured by connecting the transistor in the common base configuration illustrated in Figure 1.7(b), and plotting the collector current as a function of collector/base voltage, with the emitter current as a parameter. The resulting characteristic is shown in Figure 1.9. The common base current gain is obtained by reading off the value of collector current obtained for one of the values of emitter current and taking the ratio.

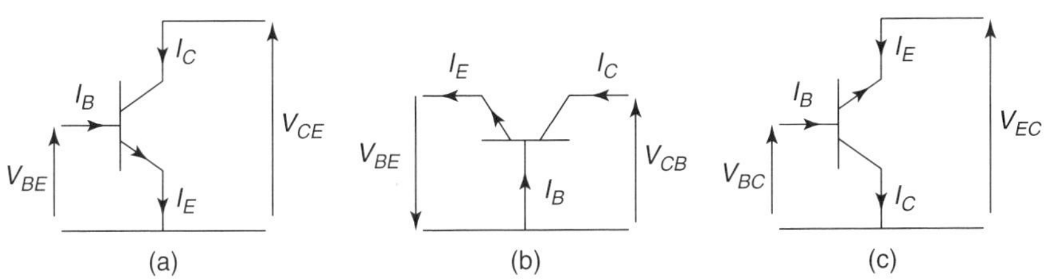

Figure 1.7 The three circuit configurations of a bipolar transistor; (a) common emitter; (b) common base; (c) common collector

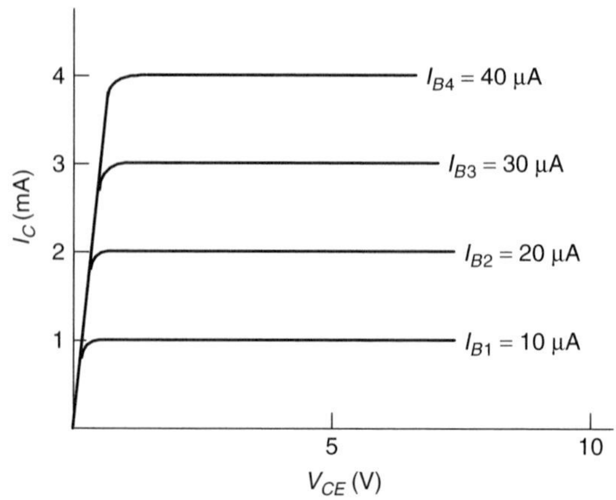

Figure 1.8 Output characteristic for a bipolar transistor connected in common emitter configuration

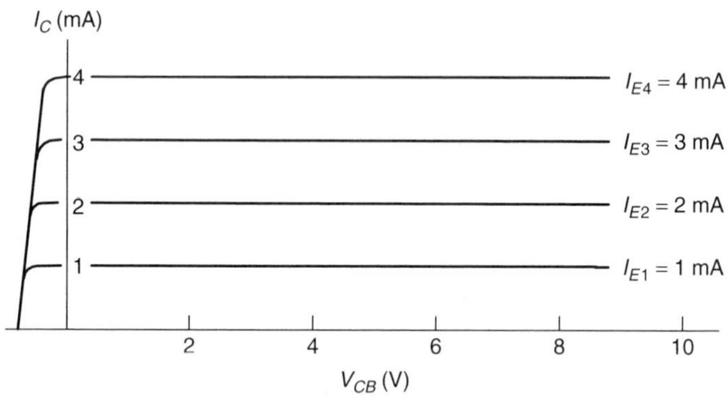

Figure 1.9 Output characteristic for a bipolar transistor connected in common base configuration

REFERENCES

[1] J. Bardeen and W.H. Brattain, 'The transistor, a semiconductor triode', *Phys. Rev.*, **74**, 230 (1948).
[2] W. Shockley, 'The theory of *p-n* junctions in semiconductors and *p-n* junction transistors', *Bell Syst. Tech. Jnl*, **28**, 435 (1949).
[3] M. Tanenbaum and D.E. Thomas, 'Diffused emitter and base silicon transistor', *Bell Syst. Tech. Jnl*, **35**, 1 (1956).
[4] H.C. Theuerer, J.J. Kleimack, H.H. Loar and H. Christenson, 'Epitaxial diffused transistors', *Proc. IRE*, **48**, 1642 (1960).

[5] J.A. Hoerni, 'Planar silicon transistor and diodes', *IRE Electron Devices Meeting*, Washington DC (1960).

[6] P. Ashburn, C.J. Bull, K.H. Nicholas and G.R. Booker, 'Effects of dislocations in silicon transistors with implanted bases', *Solid State Electronics*, **20**, 731 (1977).

[7] C. Bull, P. Ashburn, G.R. Booker and K.H. Nicholas, 'Effects of dislocations in silicon transistors with implanted emitters', *Solid State Electronics*, **22**, 95 (1979).

[8] J. Graul, H. Kaiser, W. Wilhelm and H. Ryssel, 'Bipolar high-speed, low-power gates with double implanted transistors', *IEEE Jnl Solid State Circuits*, **10**, 201 (1975).

[9] J. Graul, A. Glasl and H. Murrmann, 'High performance transistors with arsenic-implanted polysil emitters', *IEEE Jnl Solid State Circuits*, **11**, 491 (1976).

[10] T.H. Ning, R.D. Isaac, P.M. Solomon, D.D. Tang, H. Yu, G.C. Feth and S.K. Wiedmann, 'Self-aligned bipolar transistors for high-performance and low power delay VLSI', *IEEE Trans. Electron. Devices*, **28**, 1010 (1981).

[11] J. Böck, H. Knapp, K. Aufinger, M. Wurzer, S. Boguth, R. Schreiter, T.F. Meister, M. Rest, M. Ohnemus, L. Treitinger, '12 ps implanted base silicon bipolar technology', *Proc. BCTM* (1999).

[12] Special issue on Bipolar, BiCMOS/CMOS Devices & Technologies, *IEEE Trans. Electron. Devices*, **42**, No. 3 (1995).

[13] H. Higuchi, G. Kitsukawa, T. Ikeda, Y. Nishio, N. Sasaki and J. Ogiue, 'Performance and structures of scaled-down bipolar devices merged with CMOSFETS', *IEDM Technical Digest*, **694** (1984).

[14] S. Krishna, J. Kuo and I.S. Gaeta, 'An analog technology integrates bipolar, CMOS, and high voltage DMOS transistors', *IEEE Trans. Electron. Devices*, **31**, 89 (1984).

[15] K.W. Wang, P.M. Asbeck, M.F. Chang, G.J. Sullivan and D.L. Miller, 'High speed circuits for lightwave communication systems implemented with AlGaAs/GaAs heterojunction bipolar transistors', *Bipolar Circuits and Technology Meeting Digest*, **142** (1987).

[16] G.L. Paton, S.S. Iyer, S.L. Delage, S. Tiwari and J.M.C. Stork, 'Silicon germanium base heterojunction bipolar transistors by molecular beam epitaxy', *IEEE Electron. Device Lett.* **9**, 165 (1988).

[17] C.A. King, J.L. Hoyt, C.M. Gronet, J.F. Gibbons, M.P. Scott and J. Turner, 'Si/SiGe heterojunction bipolar transistors by limited reaction processing', *IEEE Electron. Device Lett.* **10**, 52 (1989).

[18] Special issue on heterostructure transistors, *IEEE Trans. Electron. Devices*, **36**, No. 10 (1989).

[19] Special issue on bipolar transistor technology; past and future trends, *IEEE Trans. Electron. Devices*, **48**, No. 11 (2001).

[20] D.L. Harame, J.H. Comfort, J.D. Cressler, E.F. Crabbé, J.Y.C. Sun, B.S. Meyerson and T. Tice, 'Si/SiGe epitaxial base transistors', *IEEE Trans. Electron. Devices*, **42**, 455 (1995).

[21] S.M. Sze, *Physics of Semiconductor Devices*, John Wiley, New York (1981).

[22] K. Hart and A. Slob, 'Integrated injection logic: a low cost bipolar logic concept', *IEEE Jnl Solid State Circuits*, **7**, 346 (1972).

2

Basic Bipolar Transistor Theory

2.1 INTRODUCTION

In this chapter a quantitative theory for the DC characteristics of a bipolar transistor is developed. The approach taken is to initially derive an approximate analytical expression for the common emitter current gain, using a simplified description of the bipolar transistor. This will allow the physical principles of the device operation to be clearly explained without resorting to undue mathematical complexity.

2.2 COMPONENTS OF BASE CURRENT

In this section the various components of base current are described. Figure 2.1 shows a schematic illustration of a bipolar transistor operating in the forward active region, that is, with the emitter/base junction forward biased and the collector/base reverse biased. This is the most common way of biasing a bipolar transistor. The forward biasing of the emitter/base junction causes electrons to be injected into the base and likewise holes into the emitter. Considering the electron current first, as electrons leave the emitter some inevitably recombine with holes in the emitter/base depletion layer. This gives rise to a recombination current I_{rg}. The remaining electrons reach the edge of the emitter/base depletion region where they become minority carriers. A concentration gradient of electrons is established in the base, which encourages them to diffuse

SiGe Heterojunction Bipolar Transistors Peter Ashburn
 ISBN: 0-470-84838-3

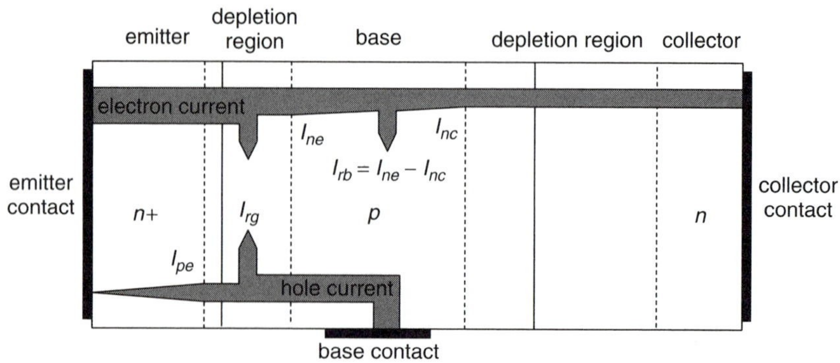

Figure 2.1 Current components in an *npn* bipolar transistor operating in the forward active mode

towards the collector. The electron diffusion current at the left-hand edge of the neutral base region is defined as I_{ne}. Further electrons recombine with holes in the base, so that the electron diffusion current at the right-hand edge of the base I_{nc} is smaller than I_{ne}. The difference between these two currents is the recombination current in the base I_{rb}. Negligible recombination occurs in the collector/base depletion region because of the high electric field across this reverse-biased junction. Similarly, once the electrons reach the n-type collector they become majority carriers, and hence no further recombination occurs.

The hole current injected from the base into the emitter is also shown in Figure 2.1. As with the electron current, a small fraction of the injected holes recombine in the emitter/base depletion region, giving rise to the recombination current I_{rg}. The remaining holes progress to the emitter where they become minority carriers, and are able to diffuse towards the emitter contact. The hole diffusion current at the left edge of the emitter/base depletion region is defined as I_{pe}. At this point two situations can arise, depending upon the thickness of the emitter W_E with respect to the hole diffusion length. If the emitter is very thick, all injected holes recombine with electrons before reaching the metal contact. In this case, the metal contact has no effect on the hole diffusion current and hence on the gain. This is the situation illustrated in Figure 2.1. Alternatively, if the emitter is very thin, the majority of the holes reach the contact without recombining. In this case, recombination occurs at the contact, and the properties of the contact have a strong influence on I_{pe}. The latter situation will be considered in the subsequent sections, since it commonly occurs in all high-speed bipolar transistors. A simple set of equations results that provide good physical insight into the physics of a bipolar transistor. In later sections, the more general situation will be considered.

By inspection of Figure 2.1 we can write the components of the emitter, collector and base currents:

$$I_E = I_{ne} + I_{rg} + I_{pe} \tag{2.1}$$

$$I_C = I_{nc} \tag{2.2}$$

$$I_B = I_E - I_C = I_{pe} + I_{rg} + I_{rb} \tag{2.3}$$

Strictly speaking, an additional current component can arise from the leakage current of the reverse-biased collector/base junction. However, in practical devices this current is of the order of $1\,nA/cm^2$ and hence can be neglected.

At this point we are in a position to define two additional parameters of the bipolar transistor. The emitter efficiency γ is defined as the ratio of the electron current injected into the base to the total emitter current:

$$\gamma = \frac{I_{ne}}{I_{ne} + I_{rg} + I_{pe}} \tag{2.4}$$

From this equation we can see that an efficient emitter is one in which I_{rg} and I_{pe} are much smaller than I_{ne}. Intuitively, we would expect I_{pe} to be smaller than I_{ne} only if the number of holes in the device was smaller than the number of electrons. This reasoning is correct, and leads to the design criterion that the emitter doping must be much larger than the base doping in order to produce an efficient emitter.

The efficiency of the base is defined by the transport factor α_T, which is the ratio of the electron current reaching the collector to that injected from the emitter:

$$\alpha_T = \frac{I_{nc}}{I_{ne}} \tag{2.5}$$

An efficient base is obtained when I_{nc} is nearly equal to I_{ne}, a situation that arises when the base is very narrow.

Finally, from equations (2.1)–(2.5) it can be seen that the common base current gain is given by:

$$\alpha = \gamma\alpha_T \tag{2.6}$$

2.3 FUNDAMENTAL EQUATIONS

The general equations for describing electron and hole transport in a semiconductor under nonequilibrium conditions are the electron and hole continuity equations:

$$\frac{\partial n}{\partial t} = G_n - U_n + \frac{1}{q}\nabla J_n \tag{2.7}$$

$$\frac{\partial p}{\partial t} = G_p - U_p - \frac{1}{q}\nabla J_p \tag{2.8}$$

where J_n and J_p are the electron and hole current densities, G_n and G_p the electron and hole generation rates ($m^{-3}s^{-1}$) due to external excitation and U_n and U_p the electron and hole recombination rates.

The solutions of these equations under appropriate boundary conditions give the electron and hole concentrations as a function of space and time. In order to arrive at an explicit solution, expressions for the current densities J_n and J_p in terms of the electron and hole concentrations are needed. These equations can readily be derived by expressing the current as the sum of a diffusion and drift term:

$$J_n = qD_n\nabla n + qn\mu_n E \tag{2.9}$$

$$J_p = -qD_p\nabla p + qp\mu_p E \tag{2.10}$$

Here the diffusion current is proportional to the gradient of the carrier concentration, indicating that carriers flow from a region of high concentration to one of low concentration. The constants D_n and D_p are the diffusion coefficients or diffusivities, and are related to the mobilities μ_n and μ_p through the Einstein relations:

$$D_n = \mu_n \frac{kT}{q} \tag{2.11}$$

$$D_p = \mu_p \frac{kT}{q} \tag{2.12}$$

In general, a further equation is needed in order to specify the electric field E. Poisson's equation provides this expression, and relates the electric field to the charge density per unit volume ρ:

$$\nabla E = \frac{\rho}{\varepsilon_0 \varepsilon_r} \tag{2.13}$$

where ε_0 is the permittivity of free space and ε_r the relative permittivity or dielectric constant. For specific problems, the charge density ρ can be expressed in terms of the electron and hole concentration, thereby providing a complete set of equations for solution.

2.3.1 Assumptions

The above equations allow a complete three-dimensional solution to be obtained for the gain of a bipolar transistor. Fortunately, however, such a rigorous analysis is not necessary, since the electrical characteristics of most practical bipolar transistors can be reasonably accurately described by a one-dimensional solution. Furthermore, a considerable simplification of the mathematics can be obtained if a number of assumptions are made.

(1) Steady-state conditions prevail, i.e.

$$\frac{\partial n}{\partial t} = \frac{\partial p}{\partial t} = 0 \tag{2.14}$$

(2) There is no external generation of carriers, i.e.

$$G_n = G_p = 0 \tag{2.15}$$

(3) All regions of the device are uniformly doped, as shown in Figure 2.2(b). This implies that there is no built-in electric field.

(4) The conductivities in the bulk semiconductor regions are high enough to ensure that all the applied voltage is dropped across the depletion regions. This assumption, when taken together with assumption 3, indicates that carriers in the bulk regions of the device move under the influence of diffusion only. This provides a considerable simplification of the mathematics, since the electric field in equations (2.9) and (2.10) can be set to zero, thereby eliminating the requirement for a solution of Poisson's equation.

(5) No generation or recombination of carriers occurs in the depletion regions of the device. This assumption is required in order that simple boundary conditions can be established for the continuity equations. It is a reasonable approximation in most circumstances, but problems can arise in some types of device, as will be discussed in Section 4.2.

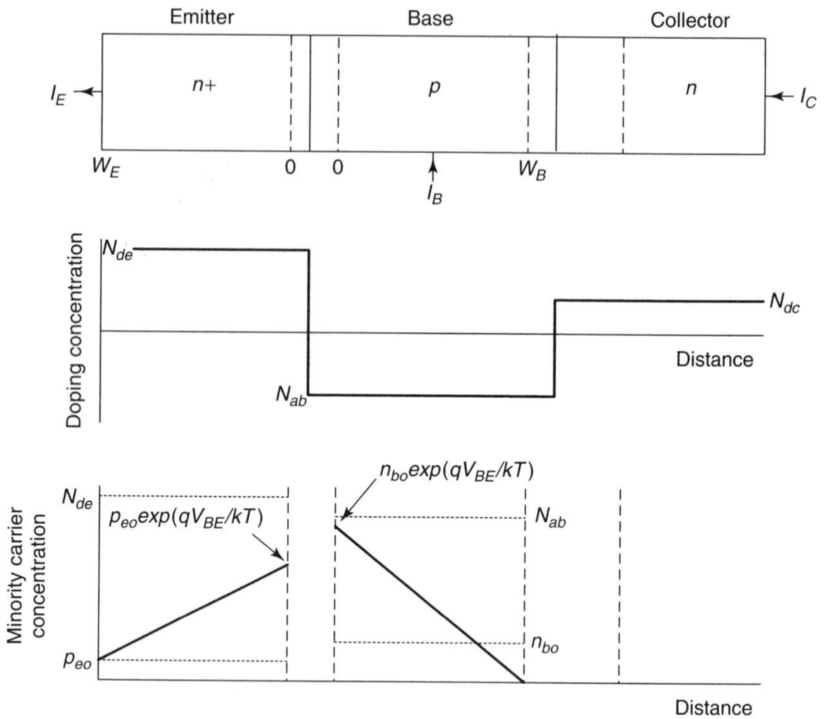

Figure 2.2 (a) One-dimensional representation of an *npn* bipolar transistor; (b) doping profiles for the case of abrupt *pn* junctions; (c) minority carrier distributions in the emitter and base for operation in the forward active region

(6) Low-level injection conditions prevail. That is, the number of electrons injected from the emitter into the base is small compared with the doping concentration in the base. This assumption is valid at low collector currents, but is violated at high currents, as will be considered in Section 4.3.

(7) The emitter is very shallow, i.e. the emitter depth W_E is less than the hole diffusion length in the emitter. The majority of minority carriers therefore diffuse across the emitter without recombining. In this case, the minority carrier distribution in the emitter is linear, as shown in Figure 2.2(c) and the properties of the emitter contact have a big influence on the current gain.

For the bipolar transistor in Figure 2.2, application of the above approximations yields simplified expressions for the electron and hole diffusion current densities:

$$J_n = qD_{nb}\frac{dn_b}{dx} \tag{2.16}$$

$$J_p = -qD_{pe}\frac{dp_e}{dx} \tag{2.17}$$

Here the subscripts b and e refer to base and emitter, respectively. The negative sign in equation (2.17) takes account of the fact that hole diffusion current flows in the direction of decreasing hole concentration. A similar procedure can also be used to simplify the continuity equations:

$$D_{nb}\frac{d^2n_b}{dx^2} - \frac{(n_b - n_{bo})}{\tau_{nb}} = 0 \tag{2.18}$$

$$D_{pe}\frac{d^2p_e}{dx^2} - \frac{(p_e - p_{eo})}{\tau_{pe}} = 0 \tag{2.19}$$

In equations (2.18) and (2.19) the recombination rates U_n and U_p have been represented by:

$$U_n = \frac{(n_b - n_{bo})}{\tau_{nb}} \tag{2.20}$$

$$U_p = \frac{(p_e - p_{eo})}{\tau_{pe}} \tag{2.21}$$

where τ_{nb} and τ_{pe} are the minority carrier lifetimes in the base and emitter, and n_{bo} and p_{eo} the thermal equilibrium values of the minority carrier concentrations in the base and emitter. The terms $(n_b - n_{bo})$ and $(p_e - p_{eo})$ therefore represent the excess minority carrier concentrations.

2.4 BASE CURRENT

The most important component of the base current in the majority of bipolar transistors is the hole diffusion current I_{pe}. This can be calculated by solving equations (2.17) and (2.19) under appropriate boundary conditions. However, for a high-speed bipolar transistor with a thin emitter, a simpler intuitive approach can be used, which only requires equation (2.17) to be solved.

2.4.1 Base Current in Shallow Emitters

If the emitter is very thin, the hole distribution in the emitter approaches a linear distribution, as illustrated in Figure 2.2. This occurs because there is little or no recombination in the bulk of the emitter when the emitter is thin, i.e. when the emitter depth W_E is much smaller than the hole diffusion length in the emitter. All holes injected from the base into the emitter recombine at the emitter contact, pinning the hole concentration at the contact to the equilibrium value i.e. p_{eo}. At the edge of the emitter/base depletion region, the hole concentration is determined by the bias across the emitter/base junction.

$$p_e(0) = p_{eo} \exp \frac{qV_{BE}}{kT} \tag{2.22}$$

In the case where all the applied voltage is dropped across the depletion region (approximation 4 in Section 2.3.1), the base current is entirely diffusion current. Equation (2.17) can then be written as:

$$J_p = -qD_{pe}(\text{gradient of } p_e) \tag{2.23}$$

The gradient of the hole distribution can be calculated from Figure 2.2

$$\text{gradient} = \frac{dp_e}{dx} = -\frac{p_{eo} \exp \dfrac{qV_{BE}}{kT} - p_{eo}}{W_E}$$

$$\therefore J_p = -qD_{pe}\frac{dp_e}{dx} = \frac{qD_{pe}p_{eo}}{W_E}\left(\exp \frac{qV_{BE}}{kT} - 1\right) \tag{2.24}$$

A more useful form of this equation can be obtained by using:

$$p_{eo}n_{eo} = n_i^2$$

$$p_{eo}N_{de} = n_i^2$$

$$p_{eo} = \frac{n_i^2}{N_{de}} \tag{2.25}$$

Substituting equation (2.25) into equation (2.24) gives:

$$I_B = \frac{qAD_{pe}n_i^2}{W_E N_{de}} \exp \frac{qV_{BE}}{kT} \tag{2.26}$$

where it has been assumed that $V_{BE} \gg kT/q$ ($V_{BE} \gg 25\,\text{mV}$ at room temperature). Equation (2.26) gives some useful insight into transistor design options, since it shows that the base current for a shallow emitter is inversely proportional to the product of the emitter depth W_E and the emitter doping N_{de}.

2.4.2 Base Current in Deep Emitters

For a deep emitter, we assume that the emitter depth is much greater than the hole diffusion length in the emitter, i.e. $W_E \gg L_{pe}$. In this case, all holes injected from the base into the emitter will recombine before they reach the emitter contact. To calculate the base current, we first have to solve equation (2.19) to determine the hole distribution in the emitter. Assuming that the emitter is infinitely thick, the boundary condition at the emitter contact is:

$$p_e(\infty) = p_{eo} \tag{2.27}$$

When this equation is used in conjunction with equation (2.22), equation (2.19) can be solved to give the following hole distribution in the emitter:

$$p_e - p_{eo} = p_{eo} \exp\frac{qV_{BE}}{kT} \exp\frac{-x}{L_{pe}} \tag{2.28}$$

where L_{pe} is the hole diffusion length in the emitter and is given by $L_{pe} = (D_{pe}\tau_{pe})^{1/2}$. Here D_{pe} is the diffusivity of holes in the emitter and τ_{pe} is the lifetime of holes in the emitter. This equation shows that the hole concentration decreases exponentially with distance in the emitter.

Combining equations (2.28) and (2.19) and solving gives:

$$I_B \approx \frac{qAD_{pe}n_i^2}{L_{pe}N_{de}} \exp\frac{qV_{BE}}{kT} \tag{2.29}$$

This equation is identical to equation (2.26) except for the replacement of W_E by L_{pe}. Equation (2.29) shows that for a deep emitter, the base current does not depend on the emitter depth W_E. This is as expected, since in a deep emitter all carriers recombine before reaching the emitter contact.

2.4.3 Recombination Current in the Neutral Base

As discussed in Section 2.2, in a transistor with a relatively wide base, there can be considerable recombination of electrons in the neutral base. This gives rise to a base recombination current. In this section, an equation for the base recombination current will be calculated. If you are reading this book for the first time, or interested purely in high-speed bipolar transistors, you can skip this section.

The recombination current in the base I_{rb} can be calculated by solving equations (2.16) and (2.18) under appropriate boundary conditions. At the base edge of the emitter/base depletion region the electron concentration is given by an equation analogous to equation (2.22):

$$n_b(0) = n_{bo} \exp \frac{qV_{BE}}{kT} \tag{2.30}$$

The second boundary condition defines the electron concentration at the base edge of the collector/base depletion region, and is given by:

$$n_b(W_B) = n_{bo} \exp -\frac{qV_{CB}}{kT} \approx 0 \tag{2.31}$$

This equation indicates that all minority carrier electrons in the vicinity of the reverse-biased collector/base junction are swept into the collector by the high electric field.

The solution of equation (2.18) with the boundary conditions in equations (2.30) and (2.31) is:

$$n_b = \frac{n_{bo} \exp \dfrac{qV_{BE}}{kT} \sinh \dfrac{W_B - x}{L_{nb}}}{\sinh \dfrac{W_B}{L_{nb}}} \tag{2.32}$$

where $L_{nb} = (D_{nb}\tau_{nb})^{1/2}$, D_{nb} is the electron diffusivity in the base and τ_{nb} is the lifetime in the base. In deriving equation (2.32), it has been assumed that $V_{BE} \gg kT/q$.

The base recombination current is given by the difference between the electron current injected into the base from the emitter and the electron current reaching the collector:

$$I_{rb} = I_{ne} - I_{nc} \tag{2.33}$$

The electron diffusion current at the edge of the emitter/base depletion region I_{ne} can be calculated from equations (2.16) and (2.32):

$$I_{ne} = qAD_{nb}\left(\frac{\mathrm{d}n_b}{\mathrm{d}x}\right)_{x=0} = -\frac{qAD_{nb}n_{bo}}{L_{nb}}\coth\frac{W_B}{L_{nb}}\exp\frac{qV_{BE}}{kT} \quad (2.34)$$

Similarly, the electron diffusion current at the edge of the collector/base depletion region I_{nc} can be calculated:

$$I_{nc} = qAD_{nb}\left(\frac{\mathrm{d}n_b}{\mathrm{d}x}\right)_{x=W_B} = -\frac{qAD_{nb}n_{bo}}{L_{nb}\sinh(W_B/L_{nb})}\exp\frac{qV_{BE}}{kT} \quad (2.35)$$

The recombination current in the neutral base is then given by:

$$I_{rb} = -\frac{qAD_{nb}n_{bo}}{L_{nb}\sinh(W_B/L_{nb})}\left(\cosh\frac{W_B}{L_{nb}} - 1\right)\exp\frac{qV_{BE}}{kT} \quad (2.36)$$

As might be expected, this equation indicates that the base recombination current is a function of the basewidth W_B of the transistor.

2.5 COLLECTOR CURRENT

In high-speed bipolar transistors, the basewidth needs to be as small as possible (typically less than 0.1 μm) so that electrons can rapidly traverse the base. The typical practical basewidth of less than 0.1 μm compares with a typical electron diffusion length in the base of 10 μm. It is clear that in this case $W_B \ll L_{nb}$ and hence that the electron distribution in the base must be linear, as shown in Figure 2.2. The collector current can therefore be calculated in a manner analogous to that used to calculate the base current in Section 2.4.1.

The electron concentration at the edge of the emitter/base depletion region is given by an equation analogous to equation (2.22):

$$n_b(0) = n_{bo}\exp\frac{qV_{BE}}{kT} \quad (2.37)$$

The electron concentration at the edge of the collector/base depletion region is given by:

$$n_b(W_B) = n_{bo}\exp-\frac{qV_{CB}}{kT} \approx 0 \quad (2.38)$$

For practical values of collector/base reverse bias the electron concentration at the edge of the collector/base depletion region is close to zero. For a linear electron distribution across the base, the diffusion equation (2.17) can then be written as:

$$J_n = qD_{nb}(\text{gradient of } n_b) = qD_{nb}\left(\frac{n_{bo}\exp\frac{qV_{BE}}{kT}}{W_B}\right) \tag{2.39}$$

$$= \frac{qD_{nb}n_{bo}}{W_B}\exp\frac{qV_{BE}}{kT} \tag{2.40}$$

A more useful form of this equation can be obtained by using:

$$p_{bo}n_{bo} = n_i^2$$

$$N_{ab}n_{bo} = n_i^2$$

$$n_{bo} = \frac{n_i^2}{N_{ab}} \tag{2.41}$$

Substituting equation (2.41) into equation (2.40) gives the collector current

$$I_C = \frac{qAD_{nb}n_i^2}{W_BN_{ab}}\exp\frac{qV_{BE}}{kT} \tag{2.42}$$

This equation gives some useful insight into transistor design options, since it shows that the collector current for a shallow emitter is inversely proportional to the product of the basewidth W_B and the base doping N_{ab}.

2.6 CURRENT GAIN

The common emitter current gain of a bipolar transistor is given by the ratio of the collector current to the base current. From equations (2.42) and (2.29) the common emitter current gain of a bipolar transistor with a shallow emitter and a thin base is given by:

$$\beta = \frac{D_{nb}W_EN_{de}}{D_{pe}W_BN_{ab}} \tag{2.43}$$

This simple equation illustrates the main design principles of a bipolar transistor. In particular, it is immediately apparent that the gain depends strongly on the ratio of the doping concentrations in the emitter and base N_{de}/N_{ab}. In order to obtain a high gain the doping concentration in the emitter should be as high as possible. Similar reasoning also suggests that the doping concentration in the base should be as low as possible. However, in practice, it is necessary to take other important electrical parameters into account. As will be explained in later chapters, the base resistance is critical in determining the switching speed of bipolar circuits, and hence it is desirable to maintain its value as low as possible. This clearly conflicts with the requirement for a high gain, and in practice an engineering compromise is arrived at, in which a gain of approximately 100 is chosen.

Equation (2.43) indicates that the common emitter current gain decreases as the emitter depth W_E decreases. This degradation of the current gain imposes a practical limit to the extent that the emitter/base junction depth can be reduced. It will be shown in later chapters that shallow emitter/base junctions are desirable in small geometry bipolar transistors in order to minimize the peripheral emitter/base capacitance. This gain degradation is therefore a serious problem in the scaling of high-speed bipolar transistors. Polysilicon emitters and heterojunction emitters, which will be discussed in Chapters 6 and 8, provide two solutions to this problem.

2.7 GUMMEL NUMBERS

The terms in equation (2.43) that relate to the base are often grouped together and called the base Gummel number:

$$G_b = \frac{W_B N_{ab}}{D_{nb}} \tag{2.44}$$

Similarly the terms that relate to the emitter can be grouped together to give the emitter Gummel number:

$$G_e = \frac{W_E N_{de}}{D_{pe}} \tag{2.45}$$

The current gain can then be written as the ratio of the emitter Gummel number to the base Gummel number:

$$\beta = \frac{G_e}{G_b} \tag{2.46}$$

3

Heavy Doping Effects

3.1 INTRODUCTION

The basic theory discussed in Chapter 2 laid the foundation for understanding the theory of operation of a bipolar transistor. However there are a number of deficiencies in the basic theory that need to be described before the theory can be applied to practical bipolar transistors. In this chapter the deficiencies of the basic theory are outlined, and a more rigorous description of the transistor behaviour produced. This requires the incorporation of additional physical mechanisms, which together are described as heavy doping effects. Heavy doping effects are difficult to model analytically, and hence as the chapter progresses increasing use will be made of empirical data that is frequently used in device simulation. This approach is entirely appropriate for the modern process and device engineer, since device simulation is an essential part of device design. The basic theory in Chapter 2 assumed uniform doping profiles for simplicity. However, in practical bipolar transistors, the profiles are rarely uniform, and hence it is necessary to consider ways of dealing with non-uniform profiles. This topic is addressed at the end of this chapter.

The simple analysis in Chapter 2 clearly indicates the desirability of using a very high doping concentration in the emitter of a bipolar transistor. Unfortunately, in reality the promised advantages of a highly doped emitter do not fully materialize. For example, the gain is significantly smaller than predicted by equation (2.43) [1], and is also strongly temperature-dependent [2]. These discrepancies between theory and experiment can be accounted for by heavy doping effects which

SiGe Heterojunction Bipolar Transistors Peter Ashburn
 ISBN: 0-470-84838-3

have not been taken into account in the simple theory. For convenience, they can be characterized by three separate but related mechanisms, namely mobility degradation at high doping concentrations [3], bandgap narrowing [4] and Auger recombination [5].

There have been many attempts to measure heavy doping effects in silicon bipolar transistors, and a number of different models have been developed. One of the difficulties in measuring heavy doping effects is that the mechanisms are interrelated. This makes it difficult to obtain unambiguous values for the model parameters. However, from the point of view of bipolar transistor modelling, provided that a consistent set of heavy doping parameters are used, the transistor characteristics can be accurately modelled. In this chapter, we will use a simple set of empirical equations for the heavy doping effects based on the work of del Alamo *et al.* [6–8]. These equations can easily be used to calculate the transistor parameters analytically and are reasonably accurate. For device simulation, a comprehensive physics-based model has been developed by Klaassen *et al.* [9–11], in which the bandgap narrowing and mobility models are unified.

3.2 MAJORITY AND MINORITY CARRIER MOBILITY

Mobility is a measure of the time interval between collisions for a carrier moving through a semiconductor lattice. The two most important collision mechanisms in bipolar transistors are lattice and impurity scattering, and the total mobility is given by the sum of the probabilities of collisions due to these individual mechanisms:

$$\frac{1}{\mu} = \frac{1}{\mu_I} + \frac{1}{\mu_L} \tag{3.1}$$

Lattice scattering is caused by collisions between carriers and the atoms of the semiconductor lattice. These lattice atoms are displaced from their lattice sites by thermal vibration, which has the effect of disrupting the perfect periodicity of the semiconductor lattice. Since thermal motion increases with temperature it is not surprising to discover that μ_L decreases with temperature. In fact it can be shown [12] that μ_L varies as $T^{-3/2}$.

Impurity scattering is caused by collisions between carriers and impurity atoms in the semiconductor lattice. As will be discussed in the following section, impurity or dopant atoms have the effect of disrupting the perfect periodicity of the semiconductor lattice, and the amount of disruption increases with impurity concentration. The mobility due to impurity scattering μ_I therefore decreases with increasing impurity concentration.

Experimental values of electron and hole mobility in silicon are shown in Figure 3.1 as a function of impurity concentration [13–15]. It can be seen that the mobility is highest at low impurity concentrations where lattice scattering is the dominant mechanism. At higher impurity concentrations both electron and hole mobilities continuously decrease with increasing dopant concentration. For silicon at impurity concentrations above $10^{19}\,cm^{-3}$ the mobilities of arsenic and phosphorus doped material are significantly different [15], as illustrated in Figure 3.2. This indicates that lower sheet resistances can be achieved with phosphorus than arsenic.

In minority carrier devices such as bipolar transistors it is the minority carrier mobility that controls the electrical characteristics of the device. In the absence of information to the contrary it has usually been assumed that the minority and majority carrier mobilities are the same. However. recent measurements of minority carrier mobility suggest that this is not the case.

A number of researchers have made measurements of minority carrier hole mobility (hole mobility in *n*-type semiconductor) in

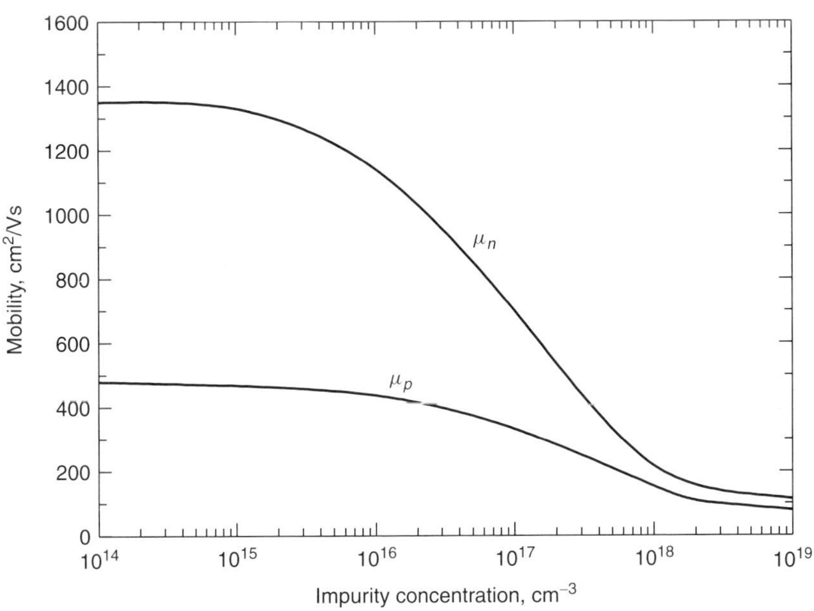

Figure 3.1 Measured values of majority carrier mobility as a function of impurity concentration for silicon (reprinted with permission from [13])

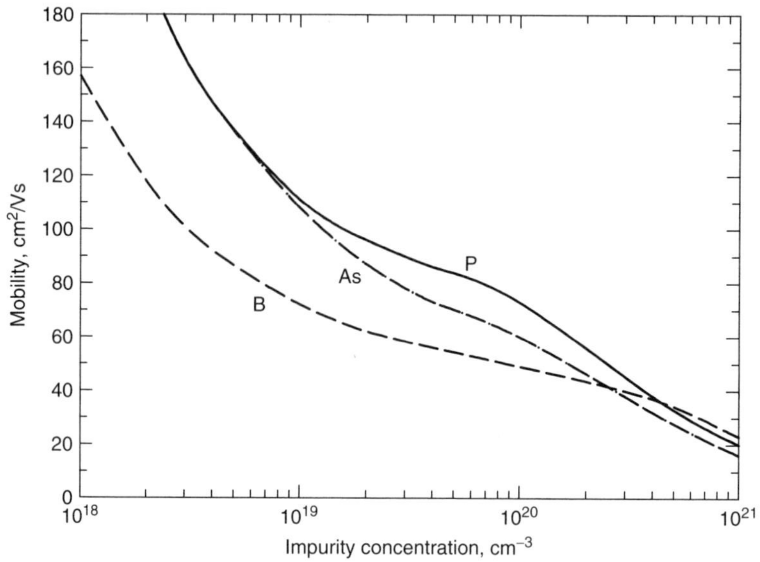

Figure 3.2 Measured values of majority carrier mobility as a function of impurity concentration in heavily doped silicon (reprinted with permission from [15])

silicon [6,8,16–18]. Figure 3.3 shows the best fit to these measured results, obtained using the following empirical equation [6]:

$$\mu_p(\min) = 130 + \frac{370}{1 + \left(\dfrac{N_D}{8 \times 10^{17}}\right)^{1.25}} \text{ cm}^2/\text{Vs} \qquad (3.2)$$

In the doping range 10^{17} to $10^{20}\,\text{cm}^{-3}$ the minority carrier mobility is over a factor of two higher than the equivalent majority carrier mobility. These experimental results are supported by theoretical calculations which predict that the minority carrier mobility is 2.8 times higher [19] at a doping concentration of around $5 \times 10^{19}\,\text{cm}^{-3}$.

Very few measurements of minority carrier electron mobility (electron mobility in *p*-type semiconductor) in silicon have been made, largely because *pnp* transistors are of less practical interest than *npn*. Figure 3.4 shows the best fit to this measured data [7] [16], obtained using the following empirical equation [7]:

$$\mu_n(\min) = 232 + \frac{1180}{1 + \left(\dfrac{N_A}{8 \times 10^{16}}\right)^{0.9}} \text{ cm}^2/\text{Vs} \qquad (3.3)$$

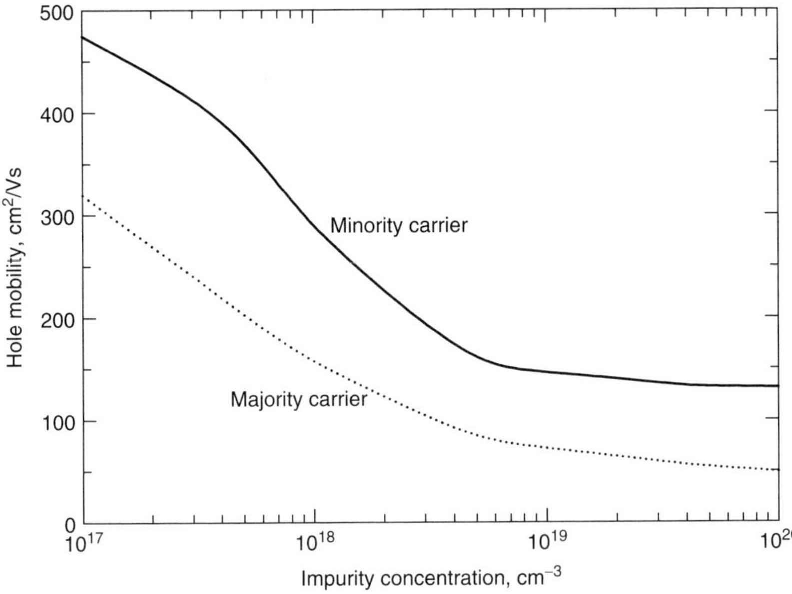

Figure 3.3 Minority carrier and majority carrier hole mobility as a function of impurity concentration in heavily doped silicon (reprinted with permission from [6])

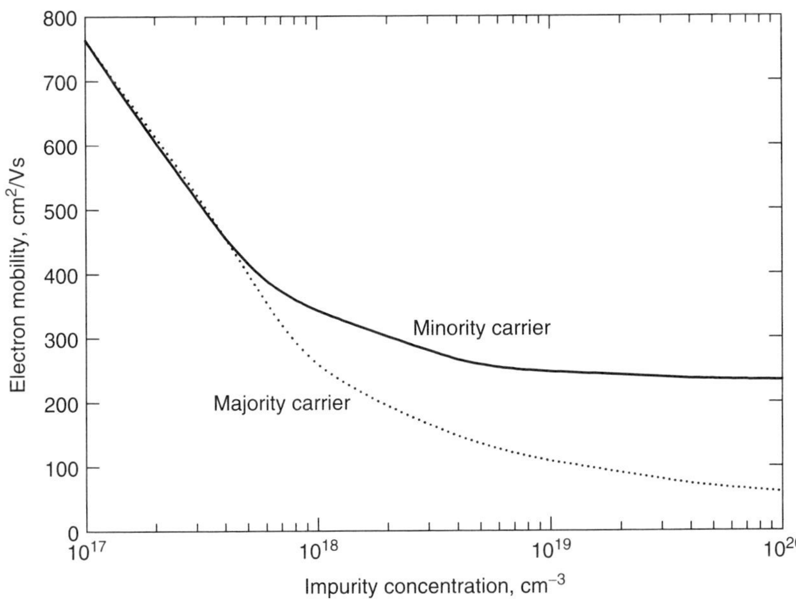

Figure 3.4 Minority carrier and majority carrier electron mobility as a function of impurity concentration in heavily doped silicon (reprinted with permission from [7])

In the doping range 10^{18} to $10^{20}\,cm^{-3}$ the minority carrier electron mobility decreases less strongly than the majority carrier mobility, and is a factor of more than two larger in heavily doped silicon.

3.3 BANDGAP NARROWING

In lightly doped semiconductors the dopant atoms are sufficiently widely spaced in the semiconductor lattice that the wave functions associated with the dopant atoms' electrons do not overlap. The energy levels of the dopant atoms are therefore discrete. Furthermore, it is reasonable to assume that the widely spaced dopant atoms have no effect on the perfect periodicity of the semiconductor lattice, and hence the edges of the conduction and valence bands are sharply defined. This situation is illustrated in the energy versus density of states diagram in Figure 3.5(a).

In heavily doped semiconductors the dopant atoms are close enough together that the wave functions of their associated electrons overlap. This causes the discrete impurity level in Figure 3.5(a) to split and form an impurity band, as shown in Figure 3.5(b). In addition, the large concentration of dopant atoms disrupts the perfect periodicity of the silicon lattice, giving rise to a band tail instead of a sharply defined band edge. Figure 3.5(b) shows the energy versus density of states diagram for the case of a heavily doped, n-type semiconductor. It can be seen that the overall effect of the high dopant concentration is to reduce the bandgap

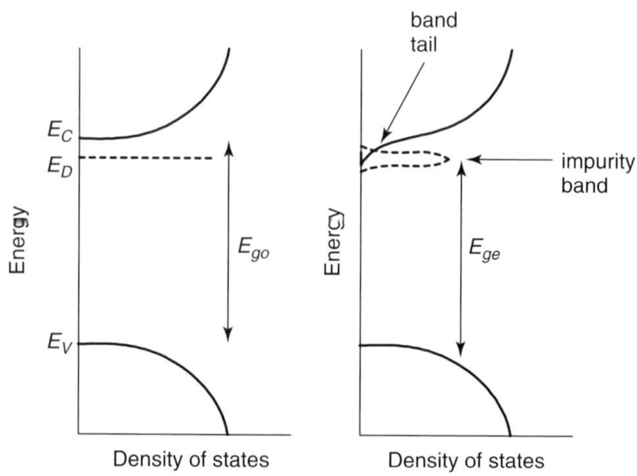

Figure 3.5 Energy versus density of states diagrams showing the effects of heavy doping on the bandgap in n-type silicon; (a) lightly doped silicon; (b) heavily doped silicon

from E_{go} to E_{ge}. A similar situation arises for a heavily doped p-type semiconductor, although in this case the bandgap narrowing occurs at the valence band edge.

At high doping concentrations, the Fermi level approaches the band edge and can even move above the band edge. In these circumstances, the Boltzmann statistics used in Chapter 2 are inaccurate and it is necessary to use Fermi-Dirac statistics to calculate the position of the Fermi level. To model heavy doping effects in the emitter of a bipolar transistor, it is necessary to combine the effects of bandgap narrowing and Fermi-Dirac statistics. For ease of modelling, these effects are rolled into a single parameter called the *apparent bandgap narrowing* or the *doping-induced bandgap narrowing* in the emitter ΔE_{ge}, which is defined by the following equation:

$$p_{eo} n_{eo} = n_{ie}^2 = n_{io}^2 \exp \frac{\Delta E_{ge}}{kT} \tag{3.4}$$

where $\Delta E_{ge} = E_{go} - E_{ge}$, n_{ie} is the intrinsic carrier concentration in the emitter, and n_{io} is the intrinsic carrier concentration for lightly doped silicon. As the name implies, the apparent bandgap narrowing is not the same as the bandgap narrowing obtained from optical measurements, because it includes the effects of Fermi-Dirac statistics. However, it can be used in combination with the equations in Chapter 2 to accurately model the electrical characteristics of a bipolar transistor.

A simple way of modelling bandgap narrowing in the emitter is through an effective doping concentration in the emitter N_{deff}:

$$N_{deff} = N_{de} \frac{n_{io}^2}{n_{ie}^2} = N_{de} \exp - \frac{\Delta E_{ge}}{kT} \tag{3.5}$$

This equation clearly indicates that bandgap narrowing has the effect of reducing the effective doping concentration in the emitter, and hence also the gain of the bipolar transistor. The gain can be calculated using equation (2.43) if the doping concentration in the emitter N_{de} is replaced by the effective doping concentration N_{deff}.

For heavily doped, n-type silicon, the model developed by del Alamo [6–8] gives a reasonably accurate description of the apparent bandgap narrowing. In this model, the apparent bandgap narrowing in the emitter ΔE_{ge} is described by the following empirical equation:

$$\Delta E_{ge} = 18.7 \ln \frac{N_{de}(\text{cm}^{-3})}{7 \times 10^{17}} \text{ meV} \tag{3.6}$$

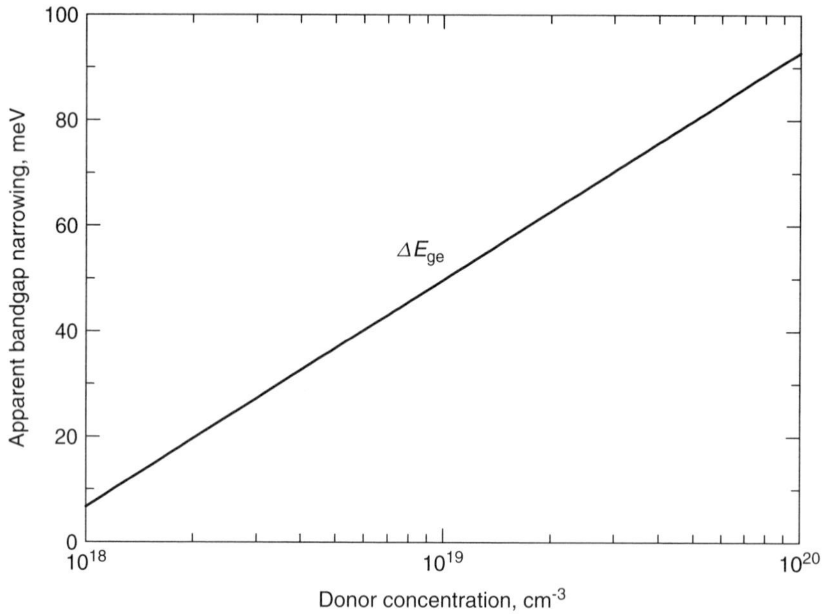

Figure 3.6 Apparent bandgap narrowing, or doping induced bandgap narrowing, as a function of donor concentration in *n*-type Si (reprinted with permission from [6])

The solid line in Figure 3.6 shows a plot of this equation. At a typical emitter doping concentration of $1 \times 10^{20}\,\text{cm}^{-3}$, the apparent bandgap narrowing is 93 meV.

In general, bandgap narrowing can also occur in a heavily doped base, or indeed in the emitter of a *pnp* transistor. This can be dealt with in an analogous way by writing the *pn* product as:

$$n_{bo}p_{bo} = n_{ib}^2 = n_{io}^2 \exp \frac{\Delta E_{gb}}{kT} \tag{3.7}$$

where ΔE_{gb} is the apparent bandgap narrowing in the base.

There have been very few measurements of apparent bandgap narrowing in *p*-type silicon, and there is some dispute about the magnitude of the effect. Swirhun *et al.* [7] proposed that the bandgap narrowing in *p*-type silicon could be described by an empirical expression originally proposed by Slotboom and de Graaff [20]:

$$\Delta E_{gb} = 9(F + \sqrt{F^2 + 0.5})\,\text{meV} \tag{3.8}$$

where

$$F = \ln \frac{N_{ab}(\text{cm}^{-3})}{10^{17}} \text{ meV} \tag{3.9}$$

The solid line in Figure 3.7 shows a plot of this equation. However, later work by Popp *et al.* [21] on the modelling of DC and AC electrical characteristics of bipolar transistors showed that equation (3.8) overestimated the bandgap narrowing and that an equation analogous to equation (3.6) gave a better fit to the measured results:

$$\Delta E_{gb} = 18.7 \ln \frac{N_{ab}(\text{cm}^{-3})}{7 \times 10^{17}} \text{ meV} \tag{3.10}$$

The dotted line in Figure 3.7 shows a plot of equation (3.10). Hence for the consistent modelling of AC and DC bipolar transistor characteristics, equation (3.10) is recommended. A comprehensive physics-based model has been developed by Klaassen *et al.* [9–11], which unifies the bandgap narrowing and mobility models, and this is recommended for device simulation.

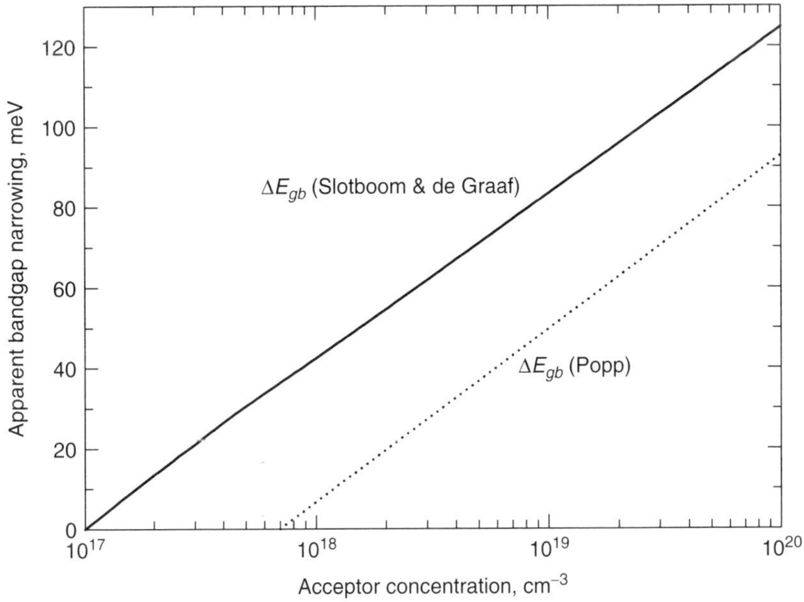

Figure 3.7 Apparent bandgap narrowing, or doping-induced bandgap narrowing [7,20,21], as a function of acceptor concentration in p-type silicon

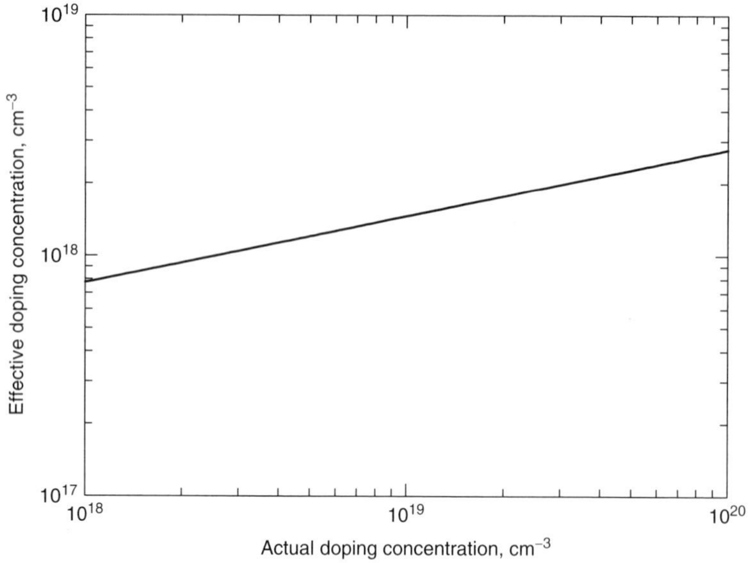

Figure 3.8 Effective doping concentration in *n*-type silicon, N_{deff}, and *p*-type silicon, N_{aeff}, as a function of actual doping concentration (after [8] and [21])

The effects of bandgap narrowing in the base can be simply modelled using an effective doping concentration in the base N_{aeff} given by:

$$N_{aeff} = N_{ab}\frac{n_{io}^2}{n_{ib}^2} = N_{ab}\exp -\frac{\Delta E_{gb}}{kT} \tag{3.11}$$

The gain can then be calculated using equation (2.43) if the doping concentration in the emitter N_{de} and in the base N_{ab} are replaced by the effective doping concentrations N_{deff} and N_{aeff}, respectively. The curve in Figure 3.8 shows a plot of both the effective doping concentrations N_{deff} and N_{aeff} in the emitter and base as a function of the actual doping concentration. It can be seen that the effective doping concentration is much lower than the actual doping concentration for doping concentrations above $1 \times 10^{18}\,\text{cm}^{-3}$.

3.4 MINORITY CARRIER LIFETIME

Experiments have shown [7,8] that the lifetime in heavily doped silicon is a strong function of doping concentration. These results can be explained by the presence of an additional recombination mechanism that is important at high doping concentrations. Auger recombination [22] has been proposed as this mechanism. This is a three-particle, band-to-band recombination mechanism, in which the energy and momentum released by the recombination of an electron-hole pair is transferred to a free electron or hole. The Auger lifetime τ_A is given by:

$$\tau_A = \frac{1}{C_N N^2} \tag{3.12}$$

where C_N is a constant known as the Auger coefficient and N is the doping concentration. The lifetime is therefore inversely proportional to the square of the doping concentration.

Experimental values of hole lifetime in n-type silicon can be fitted to an empirical equation of the form [8]:

$$\frac{1}{\tau_p} = 7.8 \times 10^{-13} \mathrm{N} + 1.8 \times 10^{-31} \mathrm{N}^2\ \mathrm{s}^{-1} \tag{3.13}$$

From equation (3.13) the Auger coefficient for holes has a value of $1.8 \times 10^{-31}\ \mathrm{cm}^6\ \mathrm{s}^{-1}$. This is in good agreement with measured values from the literature [22–24], which generally lie between 0.5 and $4.0 \times 10^{-31}\ \mathrm{cm}^6\ \mathrm{s}^{-1}$. Experimental values of electron lifetime in p-type silicon can be fitted to the following empirical equation [7]:

$$\frac{1}{\tau_n} = 3.45 \times 10^{-12} \mathrm{N} + 0.96 \times 10^{-31} \mathrm{N}^2\ \mathrm{s}^{-1} \tag{3.14}$$

The Auger coefficient for electrons therefore takes a value of $0.95 \times 10^{-31}\ \mathrm{cm}^6\ \mathrm{s}^{-1}$. Figure 3.9 shows graphs of hole and electron lifetimes as functions of doping concentration, calculated using equations (3.13) and (3.14). It can be seen that the lifetime decreases more strongly at high doping concentrations due to Auger recombination.

The hole diffusion length in the emitter L_p can be calculated from the minority carrier hole mobility μ_p in Figure 3.3 and the hole lifetime τ_p in Figure 3.9 using the following equation:

$$L_p = \sqrt{D_p \tau_p} \tag{3.15}$$

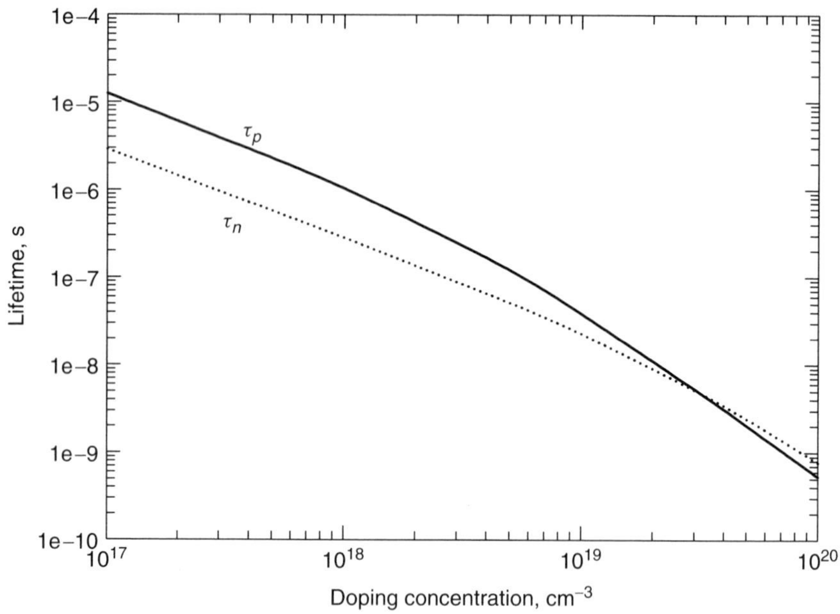

Figure 3.9 Electron lifetime τ_n in p-type silicon and hole lifetime τ_p in n-type silicon as functions of doping concentration (reprinted with permission from [6] and [7])

where D_p is the hole diffusivity given by the Einstein equation:

$$D_p = \mu_p \frac{kT}{q} \tag{3.16}$$

Similarly, the electron diffusion length in the base can be calculated from the minority carrier electron mobility μ_n in Figure 3.4 and the electron lifetime τ_n in Figure 3.9 using the following equation:

$$L_n = \sqrt{D_n \tau_n} \tag{3.17}$$

Figure 3.10 shows a graph of hole and electron diffusion length as a function of doping concentration, calculated using equations (3.15) and (3.17). For a typical emitter doping concentration of $1 \times 10^{20}\,\text{cm}^{-3}$ the hole diffusion length is around $0.4\,\mu\text{m}$, and for a typical base doping concentration of $1 \times 10^{18}\,\text{cm}^{-3}$ the electron diffusion length is around $16\,\mu\text{m}$.

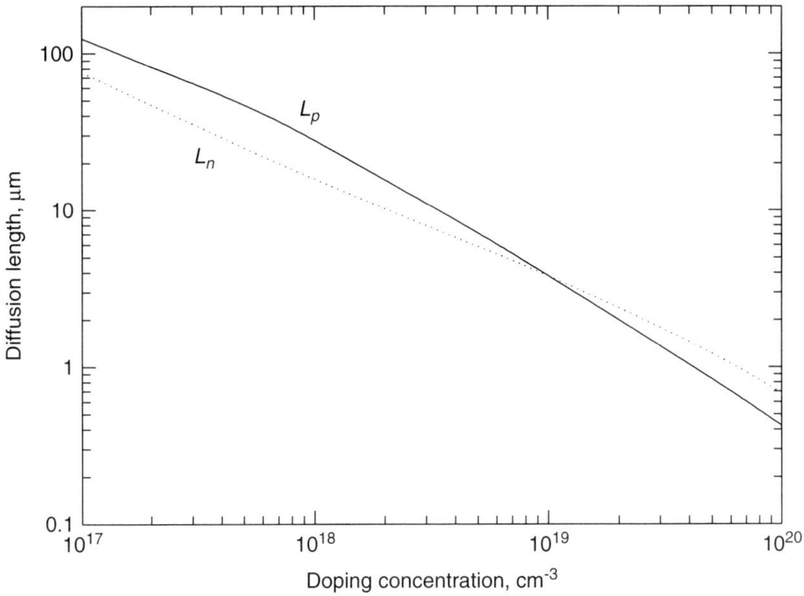

Figure 3.10 Electron diffusion length L_n in p-type silicon and hole diffusion length L_p in n-type silicon as a function of doping concentration

3.5 GAIN AND HEAVY DOPING EFFECTS

It is very simple to modify the equations for base current, collector current and gain to incorporate heavy doping effects. This is done by replacing actual doping concentrations by effective doping concentrations. The equations for base current, collector current and gain therefore become:

$$I_B = \frac{qAD_{pe}n_{io}^2}{W_E N_{deff}} \exp\frac{qV_{BE}}{kT} \tag{3.18}$$

$$I_C = \frac{qAD_{nb}n_{io}^2}{W_B N_{aeff}} \exp\frac{qV_{BE}}{kT} \tag{3.19}$$

$$\beta = \frac{D_{nb} W_E N_{deff}}{D_{pe} W_B N_{aeff}} \tag{3.20}$$

It should be noted that the intrinsic carrier concentration n_{io} used in equations (3.18) and (3.19) is the value for low-doped silicon, and D_{nb} and D_{pe} should be calculated using values of minority carrier mobility.

3.6 NON-UNIFORM DOPING PROFILES

In practical bipolar transistors, the doping concentration generally varies with depth into the silicon. In this case all the parameters, such as lifetime, mobility and bandgap narrowing, vary with depth, which makes it difficult to derive general analytical equations for the base and collector current. Device simulation is one way of dealing with this situation, which delivers accurate values of base and collector current. However, although device simulation is very effective, it does not always provide insight into the physics of the device behaviour. In this section we will therefore look for specific types of transistor where meaningful analytical solutions for the base and collector current are possible.

In the bases of bipolar transistors the doping concentration is generally relatively low (typically $2 \times 10^{18}\,\text{cm}^{-3}$), and hence Auger recombination can be ignored. The spatial variation of doping concentration, bandgap narrowing and mobility can then be incorporated with only minor alterations into the basic equations. First, equation (2.44) for the base Gummel number can be modified to take into account the non-uniform doping in the base:

$$G_b = \int_0^{W_B} \frac{N_{aeff}(x)}{D_{nb}(x)}\,\text{d}x \tag{3.21}$$

The equation for the collector current can then be written as:

$$I_C = \frac{qAn_{io}^2}{G_b} \exp \frac{qV_{BE}}{kT} \tag{3.22}$$

In the emitter of the bipolar transistor the situation is more complicated, because the doping concentration is high enough for Auger recombination to be important. Furthermore, in many cases the emitter doping is sufficiently high that the majority carriers are degenerate. In this case, Fermi-Dirac statistics must be used. These two additional factors mean that the basic theory is no longer valid, and hence a more rigorous analysis is required. Unfortunately, there is no simple means of obtaining a general analytical solution to the semiconductor equations under these conditions. However, simple and reasonably accurate solutions can be obtained in some specific types of emitter.

One group of emitters for which an analytical solution is available is shallow emitters. In these emitters, the emitter depth is small with respect to the minority carrier diffusion length, so that negligible recombination occurs in the emitter. The base current for this case has been derived by Shibib *et al.* [25]:

$$I_B = \frac{qAn_{io}^2}{\displaystyle\int_0^{W_E} \frac{N_{deff}(x)}{D_{pe}(x)}\,\mathrm{d}x + \frac{N_{deff}(W_E)}{S_M}} \exp\frac{qV_{BE}}{kT} \tag{3.23}$$

where $N_{deff}(W_E)$ is the value of the effective doping concentration at the surface of the emitter and S_M is the surface recombination velocity. For a shallow emitter with a metal contact the recombination velocity is large, and equation (3.23) can be simplified to:

$$I_B = \frac{qAn_{io}^2}{G_e(W_E)} \exp\frac{qV_{BE}}{kT} \tag{3.24}$$

$$G_e(W_E) = \int_0^{W_E} \frac{N_{deff}(x)}{D_{pe}(x)}\,\mathrm{d}x \tag{3.25}$$

where $G_e(W_E)$ is the emitter Gummel number of the transistor, which is defined in a way analogous to the base Gummel number in equation (3.21). The only unknown in equation (3.24) is the emitter doping profile $N_{de}(x)$, which can be measured using techniques such as secondary ion mass spectroscopy (SIMS). The diffusion coefficient D_{pe} and the bandgap narrowing ΔE_{ge} can be obtained from Figures 3.3 and 3.6. The base current for this type of emitter can therefore be calculated from equation (3.24) using simple numerical integration routines.

Equations (3.23)–(3.25) are valid provided that the emitter is shallow enough that the majority of recombination occurs at the surface. As the emitter depth is increased, however, an increasing fraction of minority carriers recombine in the bulk emitter, and these equations become progressively more inaccurate. A first-order correction for recombination in the emitter can be derived by modifying equation (3.23) as follows [26]:

$$I_B = \frac{qAn_{io}^2}{G_e(W_E) + \dfrac{N_{deff}(W_E)}{S_M}} \left[1 + \int_0^{W_E} \frac{[G_e(W_E) - G_e(x)]\,\mathrm{d}x}{\tau_{pe}(x)N_{deff}(x)} + \frac{N_{deff}(W_E)}{S_M}\int_0^{W_E} \frac{\mathrm{d}x}{\tau_{pe}(x)N_{deff}(x)}\right] \exp\frac{qV_{BE}}{kT} \tag{3.26}$$

where $\tau_{pe}(x)$ represents the spatial variation of the lifetime through the emitter and the term in square brackets is the correction for recombination in the emitter. This quasi-empirical equation for the base current can be solved using numerical integration routines, and is reasonably accurate for emitter/base junction depths of less than about 0.3 μm. The majority of high-speed bipolar transistors have emitter/base junction depths of less than this value and hence equation (3.26) is applicable to this type of device.

REFERENCES

[1] H.J. de Man, 'The influence of heavy doping on the emitter efficiency of a bipolar transistor', *IEEE Trans. Electron. Devices*, **18**, 833 (1971).
[2] W.L. Kauffman and A.A. Bergh, 'The temperature dependence of ideal gain in double diffused silicon transistors', *IEEE Trans. Electron. Devices*, **15**, 732 (1968).
[3] S.M. Sze and J.C. Irvin, 'Resistivity, mobility, and impurity levels in GaAs, Ge, and Si at 300 K, *Solid State Electronics*, **11**, 599 (1968).
[4] R.J. Van Overstraeten, H.J. De Man and R.P. Mertens, 'Transport equations in heavily doped silicon', *IEEE Trans. Electron. Devices*, **20**, 290 (1973).
[5] J.S. Blakemore, *Semiconductor Statistics*, Pergamon Press, Oxford (1962).
[6] J. del Alamo, S. Swirhun and R.M. Swanson, 'Simultaneous measurement of hole lifetime, hole mobility, and bandgap narrowing in heavily doped *n*-type silicon', *IEDM Technical Digest*, **290** (1985).
[7] S.E. Swirhun, Y.H. Kwark and R.M. Swanson, 'Measurement of electron lifetime, electron mobility, and bandgap narrowing in heavily doped *p*-type silicon', *IEDM Technical Digest*, **24** (1986).
[8] J. Del Alamo, S. Swirhun and R.M. Swanson, 'Measuring and modeling minority carrier transport in heavily doped silicon', *Solid State Electronics*, **28**, 47 (1985).
[9] D.B.M. Klaassen, J.W. Slotboom and H.C. de Graaff, Unified apparent bandgap narrowing in n- and p-type silicon, *Solid State Electronics*, **35**, 125 (1992).
[10] D.B.M. Klaassen, A unified mobility model for device simulation: I model equations and concentration dependence, *Solid State Electronics*, **35**, 953 (1992).
[11] D.B.M. Klaassen, A unified mobility model for device simulation: II temperature dependence of carrier mobility and lifetime, *Solid State Electronics*, **35**, 961 (1992).
[12] G.L. Pearson and J. Bardeen, 'Electrical properties of pure silicon and silicon alloys containing boron and phosphorus', *Phys. Rev.*, **75**, 865 (1949).
[13] S.M. Sze, *Physics of Semiconductor Devices*, Wiley, Chichester (1985).

[14] W.R. Thurber, R.L. Mat tis, Y.M. Liu and J.J. Filliben, 'Resistivity-dopant density relationship for boron-doped silicon', *Jnl Electrochem. Soc.* **127**, 2291 (1980).
[15] G. Masetti, M. Severi and S. Solmi, 'Modeling of carrier mobility against carrier concentration in arsenic, phosphorus, and boron doped silicon', *IEEE Trans. Electron. Devices*, **30**, 764 (1983).
[16] J. Dziewior and D. Silber, 'Minority carrier diffusion coefficients in highly-doped silicon', *App. Phys. Lett.*, **35**, 170 (1979).
[17] D.E. Burk and V. de la Torre, 'An empirical fit to minority hole mobilities', *IEEE Electron. Device Lett.*, **5**, 231 (1984).
[18] R. Mertens, J. Van Meerbergen, J. Nijs and R. Van Overstraeten, 'Measurement of the minority carrier transport parameters in heavily doped silicon', *IEEE Trans. Electron. Devices*, **27**, 949 (1980).
[19] H.S. Bennett, 'Hole and electron mobilities in heavily doped silicon: comparison of theory and experiment', *Solid State Electronics*, **26**, 1157 (1983).
[20] J.W. Slotboom and H.C. de Graaff, 'Measurement of bandgap narrowing in silicon bipolar transistors', *Solid State Electronics*, **19**, 857 (1976).
[21] J. Popp, T.F. Meister, J. Weng and H. Klose, 'Heavy doping transport parameter set describing consistently the AC and DC behaviour of bipolar transistors', *IEDM Technical Digest*, 361 (1990).
[22] J. Dziewior and W. Schmid, 'Auger coefficients for highly doped and highly excited silicon', *App. Phys. Lett.*, **31**, 346 (1977).
[23] A. Haug, 'Carrier density dependence of Auger recombination', *Solid State Electronics*, **21**, 1281 (1978).
[24] D.J. Roulston, N.D. Arora and S.G. Chamberlain, 'Modelling and measurement of minority carrier lifetime in heavy doped *N* diffused silicon diodes', *IEEE Trans. Electron. Devices*, **29**, 284 (1982).
[25] M.A. Shibib, F.A. Lindholm and F. Therez, 'Heavily doped transparent emitter regions in junction solar cells, diodes and transistors', *IEEE Trans. Electron. Devices*, **26**, 959 (1979).
[26] J.A. Del Alamo and R.M. Swanson, 'The physics and modeling of heavily doped emitters', *IEEE Trans. Electron. Devices*, **31**, 1878 (1984).

4

Second-Order Effects

4.1 INTRODUCTION

In Chapter 2, a first-order theory of bipolar transistor DC operation was described. Although this theory is valid under most circumstances, additional mechanisms come into play at the extremes of current and voltage. In this chapter, a number of second order mechanisms are described that influence the behaviour of both silicon bipolar transistors and SiGe HBTs.

The basic theory in Chapter 2 predicts that both the collector and base currents vary as $\exp(qV_{BE}/kT)$, and hence that the gain is constant. In practice however, the gain of a bipolar transistor is not constant, but varies with current, decreasing at both low and high currents, as shown in Figure 4.1(a). At low currents Figure 4.1(b) shows that the decrease in gain is due to a higher base current than expected, whereas at high currents the decrease in gain is due to a lower collector current than expected. The behaviour at low current is due to recombination in the emitter/base depletion region and at high current to high level injection. Series resistance also contributes to the behaviour at high currents. Physical explanations are given for these mechanisms and corrections to the basic equations presented. The basic theory does not predict any voltage limitation for bipolar transistor operation. In practice however, junction breakdown occurs at high voltages, which severely limits the maximum operating voltage. The physical mechanisms responsible for junction breakdown are described, along with the trade-offs involved in designing bipolar transistors for operation at a given supply voltage.

SiGe Heterojunction Bipolar Transistors Peter Ashburn

ISBN: 0-470-84838-3

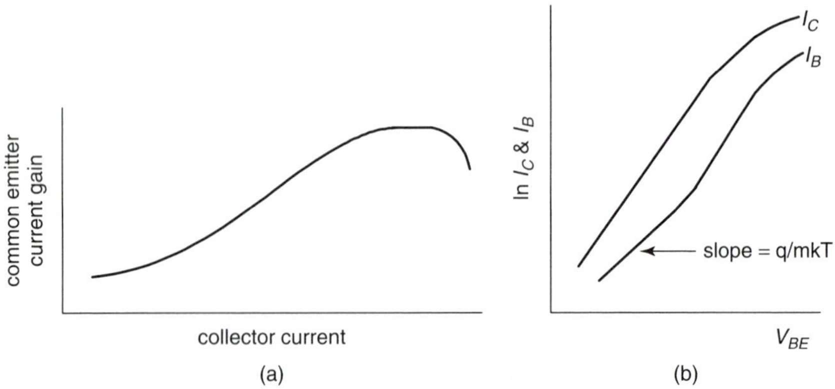

Figure 4.1 (a) Variation of common emitter current gain with collector current; (b) Gummel plot showing the origin of the variation in gain with collector current

4.2 LOW CURRENT GAIN

The decrease in gain at low currents is due to a nonideal base current, which exhibits an $\exp(qV_{BE}/mkT)$ dependence, as shown in Figure 4.1(b). The parameter m is referred to as the ideality factor, and has a value between 1 and 2. In this section it will be shown that this behaviour can be explained by recombination in the emitter/base depletion region [1], which was ignored in the basic theory (see Section 2.3.1). We will begin by considering the physics of the recombination process, and proceed to show how the simple theory can be modified to take into account recombination in the depletion region.

4.2.1 Recombination via Deep Levels

Recombination in wide bandgap semiconductors such as silicon generally occurs via deep levels located close to the centre of the bandgap. These deep levels or recombination centres arise from imperfections and impurities, and have the effect of disrupting the perfect periodicity of the semiconductor lattice. They thereby give rise to discrete energy levels in the bandgap in a similar way to donor and acceptor levels. This type of recombination is very efficient, because the deep levels act as 'stepping stones', aiding the transition of electrons and holes between the conduction and valence bands.

Figure 4.2 shows the four transitions that can occur when recombination occurs via a single deep level. The first transition is electron capture, where an electron drops from the conduction band into the

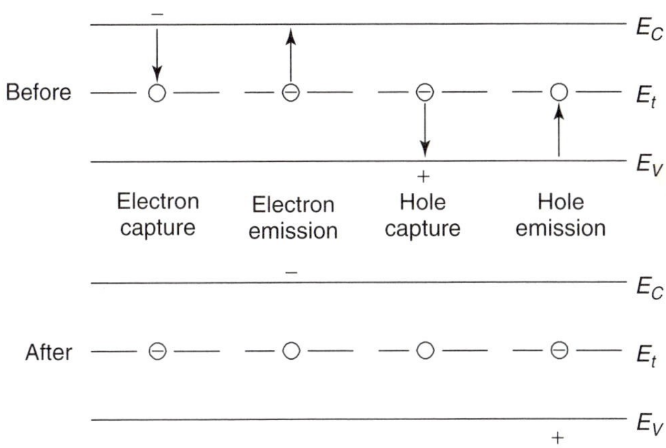

Figure 4.2 Generation/recombination processes via a deep level in the bandgap

deep level. The rate of electron capture r_{nc} is proportional to the number of electrons in the conduction band n and the number of deep levels which are not occupied by electrons p_t:

$$r_{nc} = c_n n p_t \tag{4.1}$$

The proportionality constant can be written as $c_n = v_{th}\sigma_n$, where v_{th} is the thermal velocity:

$$v_{th} = \sqrt{\frac{3kT}{m^*}} \tag{4.2}$$

The quantity σ_n is the capture cross-section for electrons and is a measure of how close the electron must come to the deep level to be captured.

The second transition in Figure 4.2 is electron emission from the deep level into the conduction band. The rate of electron emission r_{ne} is proportional to the number of deep levels that are occupied by electrons n_t:

$$r_{ne} = e_n n_t \tag{4.3}$$

The proportionality constant e_n is the electron emission probability, which can be calculated from the equilibrium case, where the net recombination $(r_{nc} - r_{ne})$ is equal to zero. In equilibrium, the electron concentration is given by equation (1.1), and the number of deep levels occupied by electrons can be written as [2]:

$$n_t = \frac{N_t}{1 + \exp{(E_t - E_F)/kT}} \tag{4.4}$$

where N_t is the total number of deep levels and E_t its energy level in the bandgap. The emission probability can therefore be calculated from equations (4.1)–(4.4):

$$e_n = v_{th}\sigma_n n_i \exp\frac{E_t - E_i}{kT} \tag{4.5}$$

This equation shows that the emission probability increases exponentially as the energy level E_t moves away from the centre of the bandgap towards conduction band. This is intuitively what we would expect, since levels close to the conduction band have a high probability of being empty of electrons.

An analogous set of equations can also be derived for hole capture and emission, namely:

$$r_{pc} = v_{th}\sigma_p p n_t \tag{4.6}$$

$$r_{pe} = p_t v_{th}\sigma_p n_i \exp\frac{E_i - E_t}{kT} \tag{4.7}$$

Under steady-state non-equilibrium conditions, such as are found in a forward-biased emitter/base junction, the rate at which electrons enter the conduction band must equal the rate at which they leave. Similarly, the rate at which holes enter the valence band must equal that at which they leave, and the net generation rate G must equal the net recombination rate U. We can then write:

$$U = r_{nc} - r_{ne} = r_{pc} - r_{pe} \tag{4.8}$$

giving:

$$U = \frac{\sigma_n\sigma_p v_{th} N_t(pn - n_i^2)}{\sigma_n\left(n + n_i\exp\frac{E_t - E_i}{kT}\right) + \sigma_p\left(p + n_i\exp\frac{E_t - E_i}{kT}\right)} \tag{4.9}$$

4.2.2 Recombination Current in the Forward Biased Emitter/Base Depletion Region

The Gummel plot in Figure 4.1 shows that there are two distinct regions to the base characteristic, which are identified by slopes of q/kT and q/mkT. The q/kT slope at high currents indicates that the basic theory of Section 2.4 applies. The base current is determined by the diffusion of holes in the emitter, and for this reason is termed diffusion current. The q/mkT slope at low currents indicates that the base current is not determined by diffusion current, and hence we need to look for a new mechanism to explain this base characteristic. We will show in this section that recombination in the emitter/base depletion region gives rise to a slope of q/mkT at low base currents.

The results in Figure 4.1 suggest that the recombination current in the emitter/base depletion region can be treated as an independent component of base current and added to the diffusion current to give the total base current. Numerical simulations [3] confirm that this assumption is correct, and indicate that the base current can be accurately described by an equation of the form:

$$I_B = I_{pe} + I_{rg} = I_1 \exp\frac{qV_{BE}}{kT} + I_2 \exp\frac{qV_{BE}}{mkT} \tag{4.10}$$

The first term models the diffusion current given by equation (3.18) and the second term the recombination current in the emitter/base depletion region.

In a forward-biased emitter/base junction the electrons lost by recombination in the depletion region give rise to a recombination current:

$$I_{rg} = qA\int_0^{W_D} U\,dx \tag{4.11}$$

where W_D is the emitter/base depletion width. Unfortunately, this equation is very difficult to solve analytically in forward bias, because the recombination rate U is a function of the electron and hole concentrations, which vary with distance across the depletion region. In this section we will take the approach of simplifying the equation for the recombination rate, and hence derive an approximate analytical equation for the recombination current.

As explained in Section 4.2.1, the most effective recombination centres lie at the centre of the bandgap. It will therefore be assumed that all the recombination centres lie at the centre of the bandgap, i.e. have an energy of $E_t = E_i$, where E_i is the intrinsic Fermi level. This is the condition for maximum recombination, and hence can be considered as the worst case. A considerable simplification of equation (4.9) results if it is also assumed that $\sigma_n = \sigma_p = \sigma$, so that:

$$U = \sigma v_{th} N_t \frac{pn - n_i^2}{n + p + 2n_i} \tag{4.12}$$

To derive an equation for the recombination current in the emitter/base depletion region, we need to obtain an expression for the pn product in equation (4.12). Equations for the electron and hole concentrations can be written in terms of the Fermi level, as shown in equations (1.1) and (1.2). However, in a forward biased pn junction, the Fermi level splits into two quasi-Fermi levels, E_{Fn} and E_{Fp}, as illustrated in Figure 2.3. In this situation, the electron and hole concentrations are given by:

$$n = n_i \exp \frac{E_{Fn} - E_i}{kT} \tag{4.13}$$

$$p = n_i \exp \frac{E_i - E_{Fp}}{kT} \tag{4.14}$$

Considerable simplifications to the mathematics arise if it is assumed that the quasi Fermi levels, E_{Fn} and E_{Fp}, are constant across the forward biased emitter/base depletion region, as illustrated in Figure 4.3. This is known as the quasi-equilibrium assumption. While this is a reasonable approximation, it is not strictly accurate in all cases. If an accurate expression for the recombination current in the emitter/base depletion

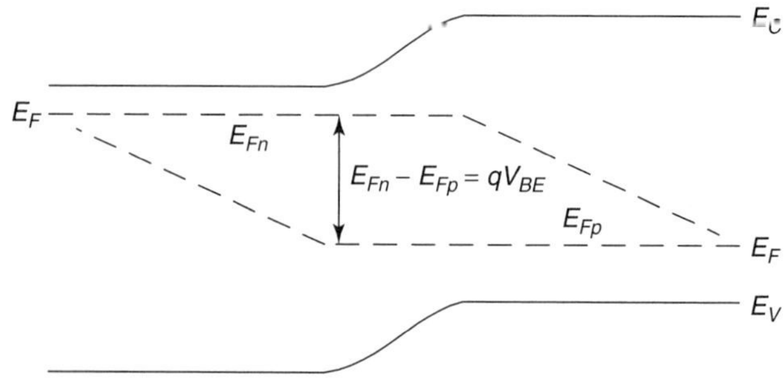

Figure 4.3 Band diagram showing the quasi-Fermi levels for an emitter/base junction under forward bias

region is needed, numerical simulation should be used [3]. The *pn* product can then be derived from equations (4.13) and (4.14):

$$pn = n_i^2 \exp \frac{qV_{BE}}{kT} \tag{4.15}$$

The recombination rate U is then given by:

$$U = \frac{\sigma v_{th} N_t n_i^2}{n + p + 2n_i} \left(\exp \frac{qV_{BE}}{kT} - 1 \right) \tag{4.16}$$

The recombination rate U is a maximum when the sum of the carrier concentrations $(n + p)$ is a minimum. It can easily be shown that the condition for a minimum is $n = p$, which occurs at the point in the depletion region where the intrinsic Fermi level is mid-way between the electron quasi-Fermi level and the hole quasi-Fermi level, as illustrated in Figure 4.4. Equations (4.13) and (4.14) can then be written as:

$$n = p = n_i \exp \frac{qV_{BE}}{2kT} \tag{4.17}$$

Substituting into equation (4.16) then gives the maximum recombination rate:

$$U_{\max} \approx \frac{1}{2} \sigma v_{th} N_t n_i \exp \frac{qV_{BE}}{2kT} \tag{4.18}$$

where it has been assumed that $V_{BE} \gg KT/q$.

An estimate for the recombination current in the emitter/base depletion region can be obtained if it is assumed that the recombination rate U

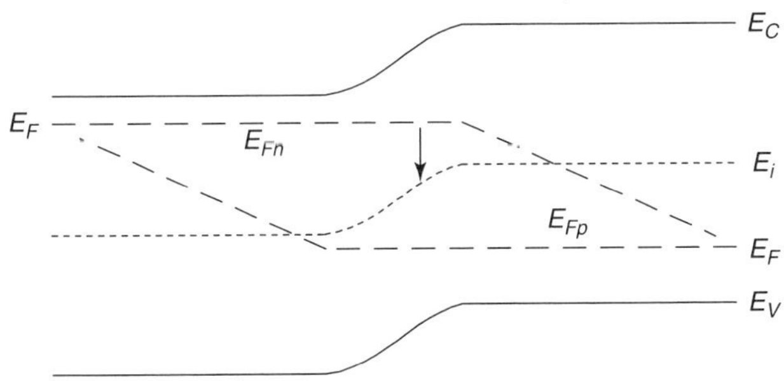

Figure 4.4 Band diagram for an emitter/base junction under forward bias showing the point in the depletion region where the intrinsic Fermi level is mid-way between the electron and hole quasi-Fermi levels

is equal its maximum value U_{max} throughout the depletion region. Evaluation of the integral in equation (4.11) is then straightforward:

$$I_{rg} = \frac{1}{2} qAW_D \sigma v_{th} N_t n_i \exp \frac{qV_{BE}}{2kT} \tag{4.19}$$

This equation essentially gives the recombination current for the worst case of strong recombination, and predicts an $\exp(qV_{BE}/2kT)$ dependence. In practical silicon bipolar transistors recombination in the emitter/base depletion region is much weaker than predicted by equation (4.19). In this case, the recombination current I_{rg} usually follows a slightly different dependence of $\exp(qV_{BE}/mkT)$, where m is the ideality factor, which takes a value between 1 and 2. Exact numerical solutions of equation (4.11) have been reported in the literature [3] and show exactly the behaviour seen in practice. This theoretical work has shown that the precise value of m depends on the physical properties of the deep level, such as its position in the bandgap, its values of electron and hole capture cross-section and the total concentration of deep levels.

4.2.3 Generation Current in a Reverse Biased *pn* Junction

The analysis in Section 4.2.2 has shown that recombination via deep levels can occur in the depletion region when a *pn* junction is forward biased. An analogous mechanism, generation via deep levels, can also occur when a junction is reverse biased. In a reverse biased *pn* junction there are very few electrons or holes, so n and p in equation (4.9) can be set to zero, giving:

$$U = \frac{\sigma v_{th} N_t n_i}{\exp \frac{E_t - E_i}{kT} + \exp \frac{E_i - E_t}{kT}} \tag{4.20}$$

$$= \frac{n_i}{\tau_0} \tag{4.21}$$

where τ_0 is a constant that depends only on the physical properties of the deep level.

The generation current in a reverse biased *pn* junction will depend on the width of the depletion region W_D and the generation rate U, and can be written as:

$$I_{gen} = qUW_D A \tag{4.22}$$

$$= q\frac{n_i}{\tau_0} W_D A \tag{4.23}$$

This equation shows that the generation current is proportional to the depletion width W_D. For an abrupt *pn* junction the depletion width varies as $V^{1/2}$ and for a linearly graded junction as $V^{1/3}$ [4]. If deep levels are present in the depletion region, we would expect generation current to dominate the reverse leakage current and hence the reverse leakage current should vary with applied voltage with a dependence between $V^{1/2}$ and $V^{1/3}$, as shown in Figure 4.5. In contrast, if no deep levels are present, diffusion current should dominate, and the reverse leakage current should be constant [4]. In practical silicon bipolar transistors and SiGe HBTs, the reverse leakage current is almost always dominated by generation in the depletion region.

4.2.4 Origins of Deep Levels in Bipolar Transistors

Generation and recombination current in bipolar transistors arises from the presence of deep levels in the silicon bandgap. One source of deep levels is the presence of impurities in the silicon. Impurities such as boron, phosphorus and arsenic in silicon introduce energy levels in the bandgap. Because these impurities are similar to silicon, the energy levels are shallow and they act as acceptor or donor atoms. Other impurities introduce energy levels that lie closer to the centre of the bandgap and hence can act as very efficient generation/recombination centres. A classic example of a deep level impurity in silicon is gold [2]. Other metals also give rise to deep levels close to the centre of the bandgap, for example copper, iron, cobalt and zinc [4]. Many of these metals have very high diffusion coefficients in silicon, and hence are able to diffuse all the way through a silicon wafer during the high-temperature processing steps required to produce a bipolar transistor. It is clear therefore that high-purity silicon wafers must be used to fabricate bipolar transistors and contact with metals avoided during transistor fabrication.

Another method by which deep levels can be introduced into the silicon bandgap is by imperfections in the silicon lattice. Ion implantation is

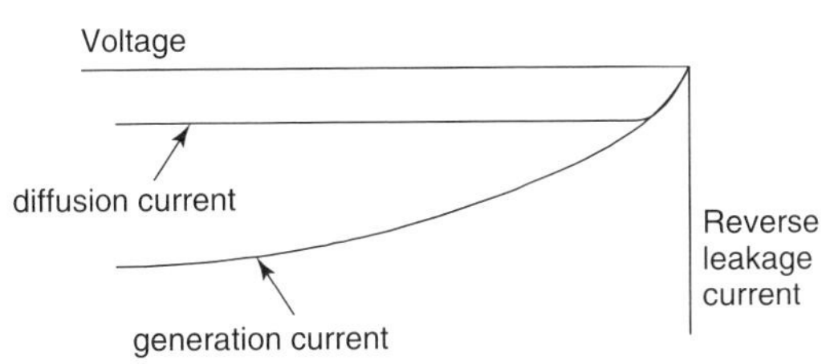

Figure 4.5 Reverse current/voltage characteristics for *pn* junctions in which generation current and diffusion current dominate

commonly used in bipolar technology, which uses high energies to place dopant atoms below the surface of the silicon. These high energy atoms introduce radiation damage into the silicon in the form of displaced atoms from their normal lattice sites. Vacancies (missing silicon atoms) and interstitials (silicon atoms on non-lattice sites) are the commonest form of radiation damage, but many more complex lattice defects also form during ion implantation. If these implantation defects are not fully removed during high-temperature annealing, deep levels will be present in the silicon that give rise to generation and recombination current. The trend in silicon technology to shallower junctions means that lower thermal budget processing will be increasingly used in bipolar transistor fabrication. Techniques such as rapid thermal annealing, which allow high-temperature anneals to be carried out for very short times, allow efficient annealing of implantation damage are very desirable in these circumstances.

The surface of the silicon wafer is another source of imperfections in the silicon lattice. These surface imperfections at the oxide/silicon interface are called surface states and have a similar effect as radiation damage in introducing deep levels into the bandgap. In bipolar transistors, the emitter/base depletion region intersects the silicon surface, where the emitter/base junction bends up to the surface at the perimeter of the emitter, as shown in Figure 4.6. It would therefore be expected that the oxide/silicon interface gives rise to generation and recombination currents in the emitter/base depletion region. The severity of the surface recombination depends on the density of surface states, which in turn depends on the quality of the oxide. High quality oxides, such as thermally grown oxides, have fewer surface states than low quality oxides, such as deposited oxides.

High-energy radiation, as is found in environments such as space, will clearly have a similar effect as ion implantation in introducing radiation damage. Generation and recombination current is one of the common

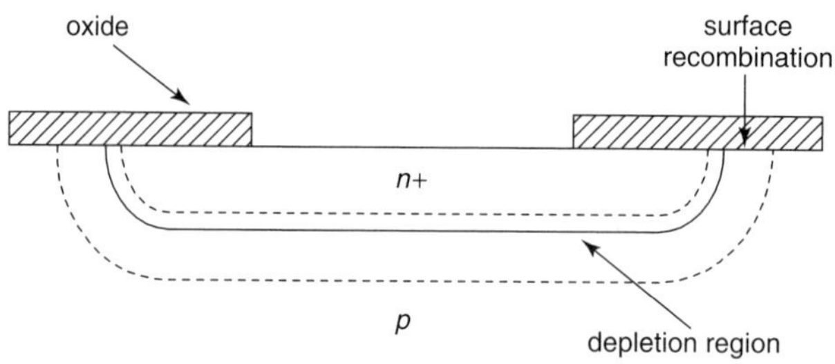

Figure 4.6 Schematic illustration of surface recombination in the emitter/base depletion region of a bipolar transistor

failure mechanisms in irradiated semiconductor devices. This failure mechanism is partly due to surface states generated at the oxide/silicon interface and partly to radiation damage in the depletion region located within the silicon.

Line and area defects, such as dislocations and stacking faults, also represent imperfections in the silicon lattice, and hence are another source of deep levels in silicon. Such defects are generally localized and hence only give rise to generation and recombination current when they intersect a depletion region. Examples of situations where line and area defects can be generated in a bipolar transistor include very heavily phosphorus-doped regions [5,6], oxidations carried out after ion implantation [7,8], and epitaxy [9]. As well as causing generation and recombination current in the depletion region, such defects also give rise to emitter-collector pipes [10,11]. Pipes are conducting paths between the emitter and collector caused by enhanced dopant diffusion along a defect that penetrates through the base, as illustrated in Figure 4.7. In SiGe HBTs, misfit dislocations can occur when relaxation of the SiGe base occurs. Such defects run parallel to the interface between the Si and the SiGe and hence are located in both the emitter/base and collector/base depletion regions, as illustrated in Figure 4.8. Misfit

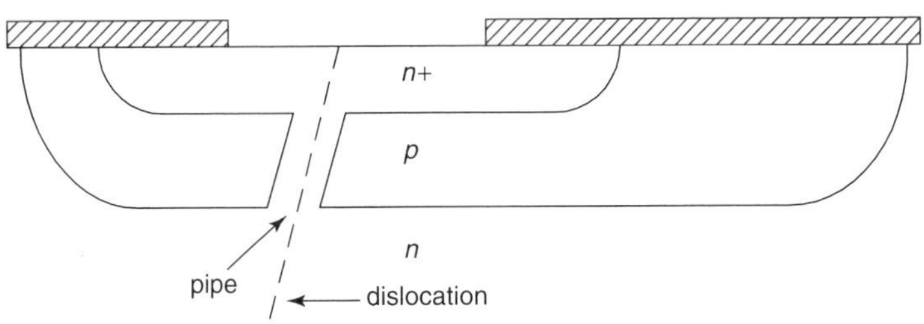

Figure 4.7 Schematic illustration of an emitter/collector pipe in a bipolar transistor, caused by enhanced diffusion along a dislocation

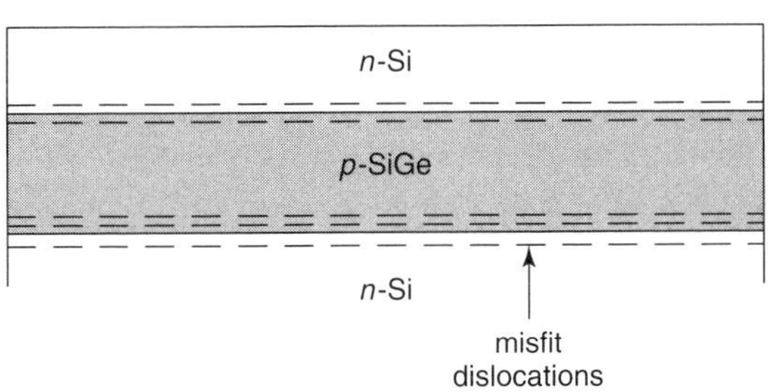

Figure 4.8 Schematic illustration showing the location of misfit dislocations in a SiGe HBT

dislocations generate deep levels in the bandgap and would be a source of generation and recombination current in the depletion regions.

4.3 HIGH CURRENT GAIN

In the simple theory of Section 2.7 it was assumed that the number of electrons injected from the emitter into the base was small with respect to the doping concentration in the base. This assumption is reasonable at moderate current levels, but at high currents the injected electron concentration may become much greater than the base doping concentration. When this occurs the hole concentration in the base must increase by the same amount as the electron concentration to maintain charge neutrality. This regime of transistor operation is referred to as conductivity modulation or high-level injection, and is illustrated in Figure 4.9. It can be seen that the electron concentration in the base, adjacent to the emitter/base depletion region, is higher than the base doping concentration. Hence the hole concentration is similarly increased above the base doping concentration to ensure overall charge neutrality. This increased hole concentration reduces both the gain and the base resistance of a bipolar transistor at high currents.

The effect of high-level injection on the Gummel plot can be physically understood by noting that the base effectively becomes more heavily doped as the injection level increases because of the extra holes that are

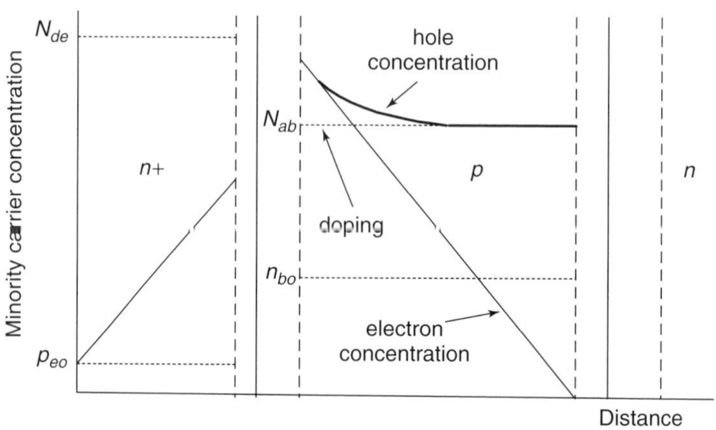

Figure 4.9 Minority carrier profiles in a bipolar transistor when the base is in the high-level injection or conductivity modulation region of operation

generated in the base to satisfy charge neutrality. Since equation (2.42) shows that the collector current is inversely proportional to the base doping concentration, high-level injection has the effect of reducing the rate of increase of the collector current with base/emitter voltage. A rigorous analysis [12] shows that the collector current varies as:

$$I_C = I_{CM} \exp \frac{qV_{BE}}{2kT} \tag{4.24}$$

Figure 4.10 shows the predicted behaviour, from which it can be seen that slope of the collector characteristic changes from q/kT to $q/2kT$ on entering high level injection. This leads to a decrease in gain at high currents. The extrapolation of the high-level injection part of the characteristic to zero base/emitter voltage gives an intercept of I_{CM}.

High-level injection is particularly important in devices in which the base doping is very low. For example, power devices such as PIN diodes or thyristors exhibit a clearly defined high-level injection region, beginning at forward voltages as low as 0.4 V [13]. In most bipolar transistors, however, the base doping is relatively high and a clearly defined transition from low- to high-level injection is hard to discern. As will be discussed later in this chapter, series resistance also causes the collector characteristic to turn over at high currents, so it is often difficult to separate these two effects in practice.

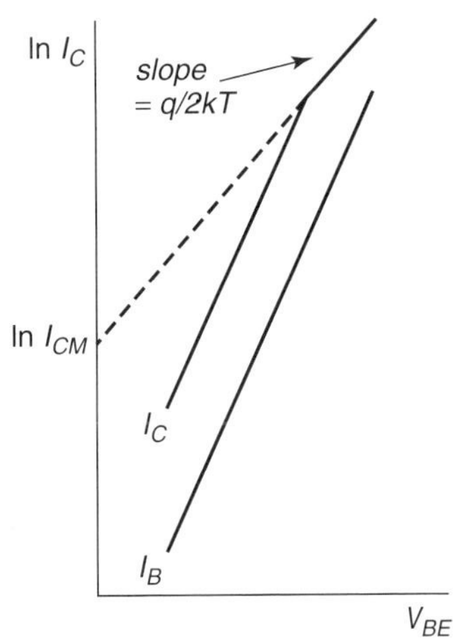

Figure 4.10 Gummel plot showing the effect of high-level injection

4.4 BASEWIDTH MODULATION

The simple theory in Section 2.4 gives no indication of how the current gain is affected by the collector/base voltage. The function of the collector/base junction in a bipolar transistor is merely to gather the minority carriers injected from the emitter into the base. We would therefore expect the collector/base voltage to have very little effect on the gain, provided, of course, that the junction does not become forward biased. Although this reasoning is broadly speaking correct, it fails to take into account the fact that increased collector/base bias gives increased penetration of the collector/base depletion region into the base. This, of course, has the effect of decreasing width of the neutral base, as illustrated in Figure 4.11.

Figure 4.11 illustrates the effect of an increase in the collector/base reverse bias on the depletion width and the minority carrier electron distribution in the base. A higher collector/base reverse bias gives a wider depletion width and hence to a narrowing of the neutral basewidth. The gradient of the injected minority carrier electron distribution in the base therefore becomes steeper. Equation (2.16) shows that the electron diffusion current is proportional to this gradient, and hence it is clear that an increase in collector/base reverse bias leads directly to an increase in

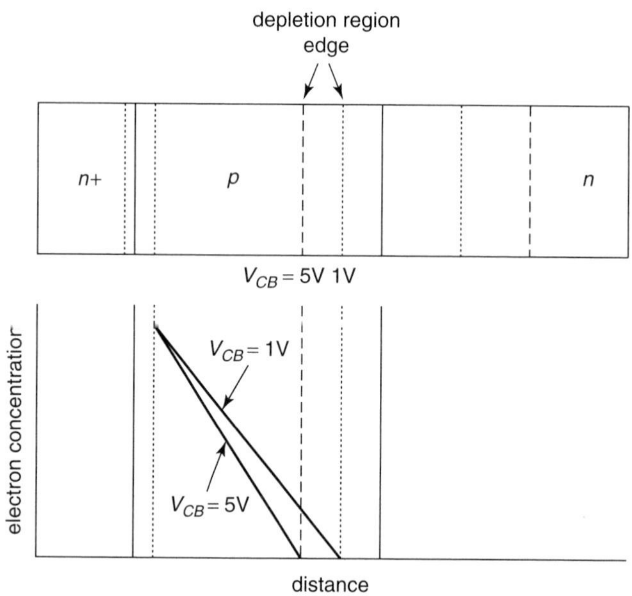

Figure 4.11 Minority carrier distribution in the base for two different values of collector/base reverse voltage

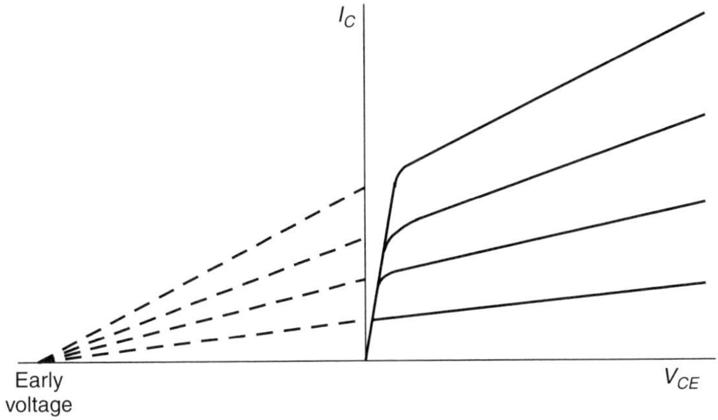

Figure 4.12 Bipolar transistor output characteristic showing the effect of basewidth modulation

collector current. This mechanism is referred to as basewidth modulation or the Early effect, and is at its strongest in transistors with a thin, lightly doped base.

Basewidth modulation influences the bipolar transistor output characteristic, as illustrated in Figure 4.12. The increase in collector current with collector/base reverse voltage is seen as a finite slope on the output characteristic. This is equivalent to a conductance at the output of the transistor, and is undesirable in many circuit applications. It is also interesting to note that if the individual characteristics in Figure 4.12 are extrapolated back along the voltage axis, they originate from a single point. This extrapolated voltage is known as the Early voltage V_{AF}, and is used as a model parameter in compact models of bipolar transistors for use in circuit simulators, as will be discussed in Chapter 11.

4.5 SERIES RESISTANCE

In practical bipolar transistors, the silicon that is used to create the emitter, base and collector of the bipolar transistor has some series resistance. We would therefore expect series emitter resistance, base resistance and collector resistance to limit the current that the bipolar transistor can deliver. In a silicon bipolar transistor, the emitter is generally heavily doped, the base moderately doped and the collector lightly doped. We would therefore expect collector resistance to be very high, base resistance moderately high and emitter resistance small. This is indeed the case in practice, as will be described in Chapter 5.

Collector resistance is a particular problem, and special processing techniques (buried layer and epitaxy) have been developed to reduce the collector resistance, as will be described in Chapter 9. Minimization of base resistance is also vitally important, since it has a strong influence on the switching speed of bipolar circuits, as will be discussed in Chapter 12.

The influence of series resistance on the transistor currents can be understood from the circuit diagram in Figure 4.13. The external connections to the transistor are the terminals C, B and E, whereas the internal terminals of the ideal transistor that we have been discussing so far are the terminals C′, B′ and E′. Of course there is no way of gaining access to these internal terminals of the transistor in practice. The relationship between the internal and external base/emitter voltages can be found using Kirchoff's voltage law:

$$\begin{aligned} V_{B'E'} &= V_{BE} - I_B R_B - I_E R_E \\ &= V_{BE} - I_B R_B - (I_C + I_B) R_E \\ &= V_{BE} - I_B R_B - I_B R_E (1 + \beta) \end{aligned} \tag{4.25}$$

The collector current is then given by:

$$I_C = I_S \exp \frac{qV_{B'E'}}{kT} = I_S \exp \frac{q\,(V_{BE} - I_B R_B - I_B R_E(1+\beta))}{kT} \tag{4.26}$$

Equation (4.26) shows that at low currents, the external and internal base/emitter voltages will be approximately the same, so the collector

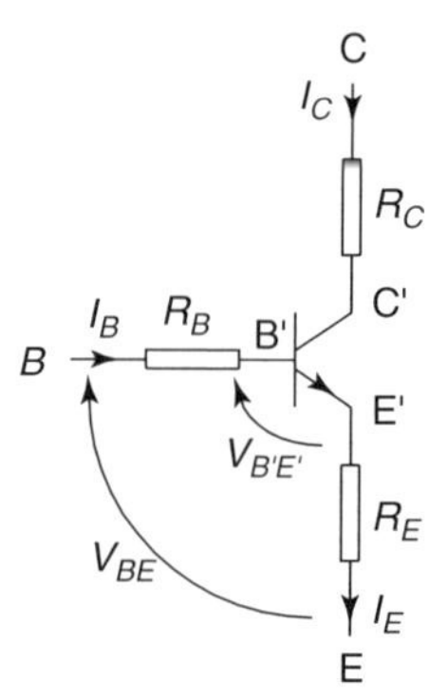

Figure 4.13 Circuit diagram showing internal collector, base and emitter series resistances

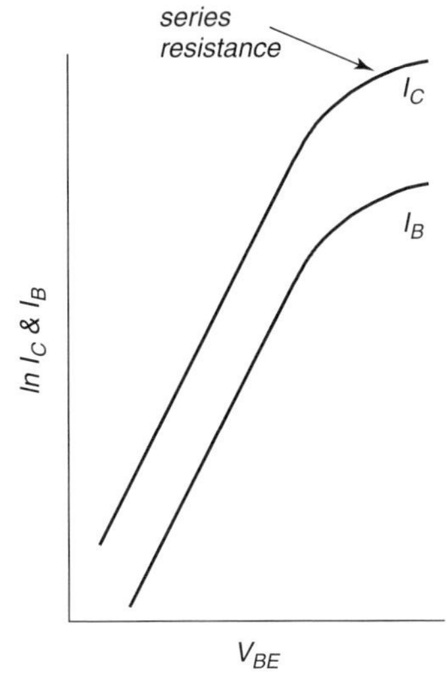

Figure 4.14 Gummel plot showing the effect of series resistance at high current

current will be given by the basic theory; i.e. equation (3.19). However, at high currents, the voltage drop across the base and emitter resistance will cause the internal base/emitter voltage to be smaller than the external base/emitter voltage, with the result that the collector current will be smaller than predicted by the basic theory. The net result is that the collector characteristic will turn over at high currents, as illustrated in Figure 4.14. Similar behaviour is seen at high currents in the base characteristic. It should be noted that, although the emitter resistance is generally very small, it is multiplied by the gain of the bipolar transistor in equation (4.26). Minimization of emitter resistance, as well as base and collector resistance, is therefore important in the design of bipolar transistors. Further information on the origin of base, collector and emitter resistance is given in Chapter 5.

4.6 JUNCTION BREAKDOWN

There is a limit to the reverse voltage that can be applied to the collector of a bipolar transistor. At high reverse voltages, the junction breaks down and a high current flows between the emitter and collector. The voltage at which this occurs is known as the breakdown voltage. No transistor action is obtained above the breakdown voltage, and hence this imposes an upper limit on the supply voltage of the circuit in which the transistor is used.

It is also interesting to note that a lower breakdown voltage is obtained when the transistor is connected in the common emitter mode than the common base mode. At first sight, this is somewhat surprising, since in both cases it is the collector/base junction that is breaking down. In Section 4.6.5 we will explain how the current gain of the transistor is responsible for this difference.

Several physical mechanisms can give rise to excessive current at high collector voltages, the most important of which are punch-through, Zener breakdown and avalanche breakdown. The first two mechanisms can usually be avoided by careful transistor design, but avalanche breakdown imposes a fundamental limit on the operating voltage of bipolar transistors.

4.6.1 Punch-through

In Section 4.4 it was shown how the application of a reverse bias to the collector caused the collector/base depletion region to extend into the base and hence modulate the basewidth. In the limit, the application of a reverse bias to the collector could cause the depletion region to extend across the whole width of the base and join up with the emitter/base depletion region. The emitter and collector are then connected together by a single depletion region, as illustrated in Figure 4.15. This is known as punch-through, and when it occurs a large current flows between emitter and collector. Its electrical effect is similar to junction breakdown, although, of course, the physical mechanism is completely different.

State-of-the-art, silicon bipolar transistors typically have basewidths of much less than 0.1 μm, and consequently often operate close to the punch-through limit. Careful transistor and process design is therefore required in order to ensure that punch-through does not occur.

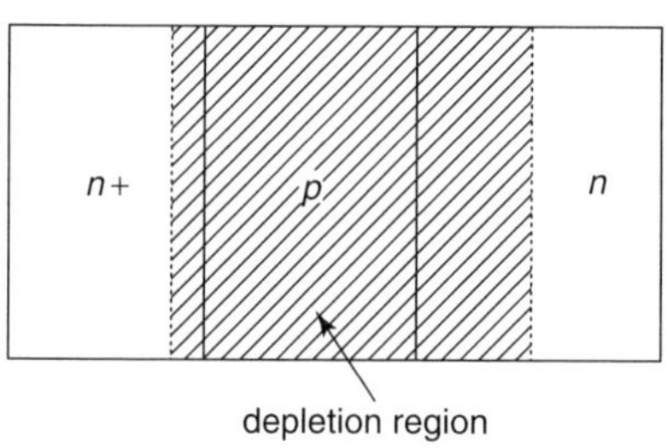

Figure 4.15 Schematic illustration of a bipolar transistor operating in punch-through

From these considerations it is also clear that punch-through imposes a fundamental limit to the scaling of the basewidth of a bipolar transistor.

4.6.2 Zener Breakdown

Zener breakdown is a tunnelling mechanism in which large numbers of carriers penetrate through the energy barrier imposed by the bandgap of the semiconductor. This is illustrated schematically in Figure 4.16 for a reverse-biased *pn* junction. For tunnelling to occur, the barrier presented to the tunnelling carriers must be very thin. This situation only arises at electric fields above approximately 10^6 V/cm. In general, such high electric fields only occur when both the *n* and *p* regions are very heavily doped. In practical transistors, tunnelling is therefore most likely to be seen in the reverse emitter/base diode characteristics [14,15].

The tunnelling mechanism illustrated in Figure 4.16 is referred to as band-to-band tunnelling, since carriers tunnel from one band directly to another. Band-to-band tunnelling can be described by an equation of the form [15,16]:

$$J_{bbt} = c_{bbt} V_{bi} F_m^{3/2} \exp\left(-\frac{F_0}{F_m}\right) \tag{4.27}$$

$$c_{bbt} = cq \tag{4.28}$$

where V_{bi} is the built-in junction voltage, F_m is the maximum electric field and the other parameters are constants with the following values:

$$F_0 = 1.93 \times 10^9 \text{ V/m}$$

$$c = 5 \times 10^{15} \text{ cm}^{-1/2}\text{ V}^{-5/2}\text{ s}^{-1} \text{ (best case) to}$$

$$4 \times 10^{17} \text{ cm}^{-1/2}\text{ V}^{-5/2}\text{ s}^{-1} \text{ (worst case)}$$

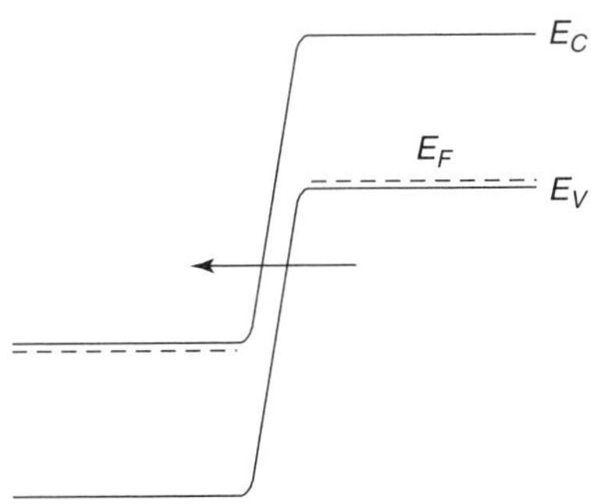

Figure 4.16 Band diagram illustrating the mechanism of Zener breakdown

Tunnelling can also occur via traps. In this case, it is referred to as trap-assisted tunnelling, and can be described by an equation of the form [15,17]:

$$J_{tat} = \sqrt{3\pi}\frac{qn_i}{2\tau_g}W\frac{\gamma}{F}\left[\exp\left(\frac{F}{\gamma}\right)^2 - \exp\left(\frac{FW_0}{\gamma W}\right)^2\right] \tag{4.29}$$

$$\gamma = \frac{\sqrt{24m^*(kT)^3}}{q\hbar} \tag{4.30}$$

where τ_g is the generation lifetime, W is the depletion width, W_0 is the zero bias depletion width, and F is the electric field.

4.6.3 Avalanche Breakdown

Avalanche multiplication or impact ionization is by far the most common breakdown mechanism in practical bipolar transistors. In a reverse-biased *pn* junction, electron–hole pairs are continually being generated by thermal agitation. At low reverse voltages this gives rise to a leakage generation current, which can be calculated from equation (4.23). At high reverse voltages, however, the generated carriers gain sufficient kinetic energy between collisions with the silicon lattice for them to be able to shatter the silicon-silicon bond. This mechanism is referred to as impact ionization, and leads to the generation of an electron–hole pair. The original carrier and the electron and hole generated are then accelerated in opposite directions by the electric field, and in turn are able to produce further electron–hole pairs by impact ionization. This process, known as avalanche multiplication, rapidly leads to the generation of large numbers of carriers and hence to a large current.

For avalanche multiplication to occur, a critical electric field E_{crit} must be established across the reverse-biased junction. Since the depletion width depends upon the doping concentration it is clear that the breakdown voltage BV will also depend on the doping concentration. For a one-sided step junction the breakdown voltage is given by [4]:

$$BV = \frac{\varepsilon_0\varepsilon_r E_{crit}^2}{2qN_L} \tag{4.31}$$

where N_L is the doping concentration on the lightly doped side of the junction. If E_{crit} was a constant, equation (4.31) would indicate that the breakdown voltage was inversely proportional to the doping concentration. In practice, however, E_{crit} varies slightly with doping concentration [4], taking values between 3×10^5 and 1×10^6 V/cm [19].

4.6.4 Junction Breakdown in Practice

In practical reverse-biased emitter/base diode characteristics, a mixture of mechanisms is generally seen, with different mechanisms dominating different parts of the reverse characteristic, as illustrated in Figure 4.17. Avalanche breakdown dominates at high reverse-bias, band-to-band tunnelling (equation (4.27)) just prior to avalanche breakdown and a mixture of generation via deep levels (equation (4.22)) and trap-assisted tunnelling (equation (4.29)) at low bias. An effective method of distinguishing between Zener breakdown and avalanche breakdown is by measuring the temperature dependence of the breakdown voltage. If Zener breakdown is the dominant mechanism, the breakdown voltage decreases with increasing temperature, whereas for avalanche breakdown it increases with temperature [18].

4.6.5 Common Base and Common Emitter Breakdown Voltages

In bipolar transistors, the breakdown voltage depends on the way that the bipolar transistor is connected in the circuit. In common base connection, the breakdown voltage obtained is the same as that predicted by equation (4.31), whereas in common emitter connection the

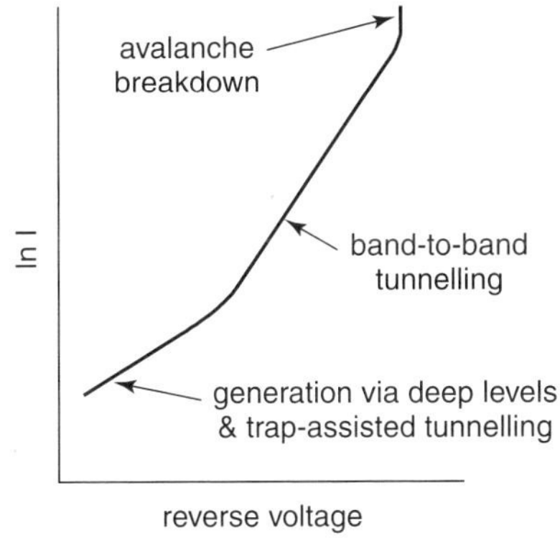

Figure 4.17 Schematic illustration of a reverse emitter/base diode characteristic showing the mechanisms dominating the different parts of the characteristic

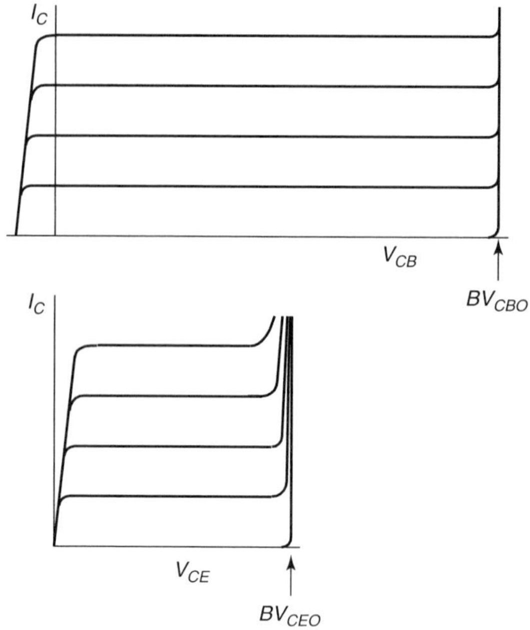

Figure 4.18 Output characteristics and breakdown voltages of a bipolar transistor connected in common base (top) and common emitter (bottom) configurations

breakdown voltage is considerably lower, as illustrated in Figure 4.18. In practice, the breakdown voltage in bipolar transistors is measured with the base open circuit, and hence in common base mode the breakdown voltage is referred to as BV_{CBO} (breakdown voltage in common base connection with the base open circuit). In the common emitter mode the breakdown voltage is referred to as BV_{CEO} (breakdown voltage in common emitter connection with the base open circuit).

The lower breakdown voltage in common emitter connection can be understood by considering the currents flowing in the transistor when it is connected in common emitter configuration. With reference to Figure 4.19, if the current flowing across the emitter/base junction is I_F, a fraction of this current is collected at the collector/base junction, given by αI_F, where α is the common base current gain, as described in Chapter 1. In addition, there will be a component of current at the collector due to the leakage current of the collector/base junction I_{CBO}. In this case, we can write:

$$I_E = I_F \tag{4.32}$$

$$I_C = \alpha I_F + I_{CBO} \tag{4.33}$$

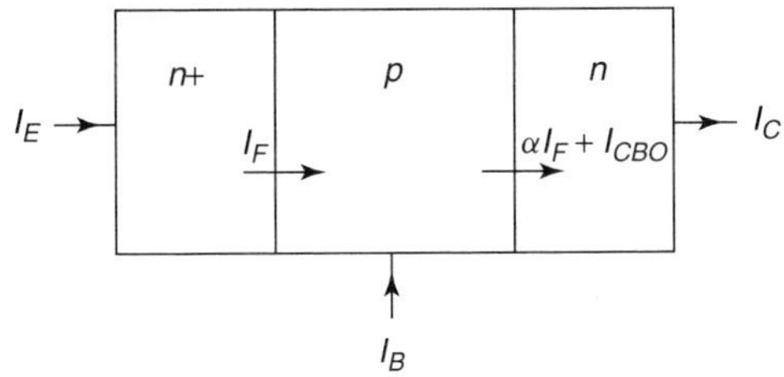

Figure 4.19 Schematic illustration showing the components of current flowing across the emitter/base and collector/base junctions

When the collector/base junction is breaking down, the current across the junction is multiplied by the electron–hole pairs created by avalanche breakdown. This junction breakdown can be modelled using an empirical expression for the multiplication factor M [20]:

$$M = \frac{1}{1 - \left(\frac{V_{CB}}{BV_{CBO}}\right)^n} \tag{4.34}$$

where n takes a value between 3 and 6. In this case, the current at the collector/base junction is multiplied by M:

$$I_C = M(\alpha I_F + I_{CBO}) \tag{4.35}$$

If the base is open circuit, the emitter current must equal the collector current, and therefore equations (4.32) and (4.35) can be equated:

$$I_C = I_E = I_{CEO} = M(\alpha I_{CEO} + I_{CBO})$$
$$\therefore I_{CEO} = \frac{MI_{CBO}}{1 - \alpha M} \tag{4.36}$$

where I_{CEO} is the current flowing between emitter and collector when the base is open circuit. Equation (4.36) shows that the collector/emitter current begins to increase very rapidly when αM approaches unity. In contrast, in the common base mode the collector/base leakage current only begins to increase when αM approaches infinity. This explains why the breakdown voltage in the common emitter mode BV_{CEO} is lower than that in the common base mode BV_{CBO}.

The value of BV_{CEO} can be calculated by noting that when breakdown occurs, $\alpha M = 1$ and V_{CB} is equal to BV_{CEO}. Using equation (4.34) gives:

$$\frac{\alpha}{1-\left(\dfrac{BV_{CEO}}{BV_{CBO}}\right)^n} = 1$$

$$\therefore BV_{CEO} = BV_{CBO}(1-\alpha)^{1/n} = \frac{BV_{CBO}}{\beta^{1/n}} \tag{4.37}$$

where equation (1.7) has been used.

4.6.6 Trade-off between Gain and BV_{CEO}

Equation (4.37) shows that the common emitter breakdown voltage BV_{CEO} is inversely proportional to the common emitter current gain of the transistor. There is therefore a trade-off between gain and breakdown voltage. Clearly a high gain and a high breakdown voltage cannot be obtained simultaneously, so a compromise must be reached between reasonable values of gain and breakdown voltage. Figure 4.20 shows a plot of equation (4.37), where $n = 5$ and $BV_{CBO} = 7\,\text{V}$ have been used. It can be seen that a gain of 100 gives a common emitter breakdown

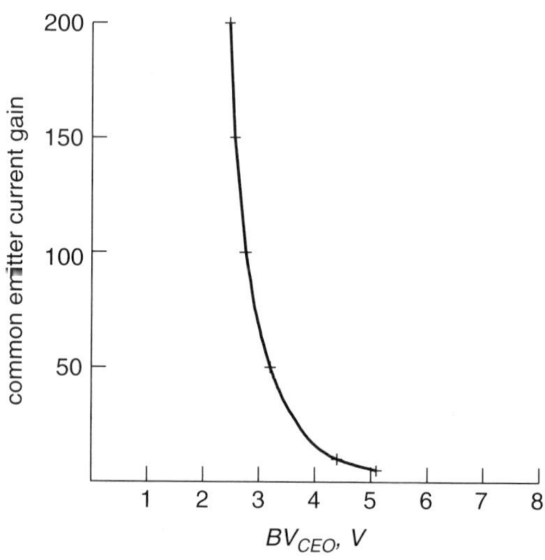

Figure 4.20 Graph showing the trade-off between common emitter current gain β and common emitter breakdown voltage BV_{CEO}

voltage BV_{CEO} of 2.8 V, which is considerably lower than the common base breakdown voltage BV_{CBO} of 7 V.

REFERENCES

[1] C.T. Sah, R.N. Noyce and W. Shockley, 'Carrier generation and recombination in *pn* junctions and *pn* junction characteristics', *Proc. IRE*, **45**, 1228 (1957).

[2] A.S. Grove, *Physics and Technology of Semiconductor Devices*, Wiley, Chichester (1971).

[3] P. Ashburn, D.V. Morgan and M.J. Howes, 'A theoretical and experimental study of recombination in silicon *p-n* junctions', *Solid State Electronics*, **18**, 569 (1975).

[4] S.M. Sze, *Physics of Semiconductor Devices*, Wiley, Chichester (1981).

[5] C.J. Bull and P. Ashburn, 'A study of diffused bipolar transistors by electron microscopy', *Solid State Electronics*, **23**, 953 (1980).

[6] P. Ashburn and C.J. Bull, 'Observations of dislocations and junction irregularities in bipolar transistors using the EBIC mode of the scanning electron microscope', *Solid State Electronics*, **22** 105 (1979).

[7] P. Ashburn, C. Bull, K.H. Nicholas and G.R. Booker, 'Effects of dislocations in silicon transistors with implanted bases', *Solid State Electronics*, **20**, 731 (1977).

[8] C. Bull, P. Ashburn, G.R. Booker and K.H. Nicholas, 'Effects of dislocations in silicon transistors with implanted emitters', *Solid State Electronics*, **22**, 95 (1979).

[9] K.V. Ravi, '*Imperfections and impurities in semiconductor silicon*', Wiley, Chichester (1981)

[10] J.P. Gowers, C.J. Bull and P. Ashburn, 'SEM and TEM observations of emitter/collector pipes in bipolar transistors', *Jnl Microscopy*, **118**, 329 (1980).

[11] P. Ashburn, C.J. Bull and J.R.A. Beale, 'The use of electron beam induced current mode of the SEM for observing emitter/collector pipes in bipolar transistors', *Jnl App. Phys.*, **50**, 3472 (1979).

[12] R.N. Hall, 'Power rectifiers and transistors', *Proc. IRE*, **44**, 72 (1956).

[13] B.W. Wessels and B.I. Baliga, 'Vertical channel field controlled thyristors with high gain and fast switching speeds', *IEEE Trans. Electron. Devices*, **25**, 1261 (1978).

[14]. A. Cuthbertson and P. Ashburn, 'Self-aligned transistors with polysilicon emitters for bipolar VLSI', *IEEE Trans. Electron. Devices*, **32**, 242 (1985).

[15] G.A.M. Hurkx, H.C. de Graaff, W.J. Kloosterman and M.P.G. Knuvers, 'A novel compact model description of reverse biased diode characteristics including tunnelling', *Proc. ESSDERC*, 49 (1990).

[16] E.O. Kane, 'Theory of tunnelling', *Jnl. Appl. Phys.* **32**, 83 (1961).

[17] T.F. Meister, J. Schäfer, M. Franosch, W. Molzer, K. Aufinger, U. Scheler, C. Walz, M. Stolz, S. Boguth and J. Böck, 'SiGe base bipolar technology with 74 GHz f_{max} and 11 ps gate delay', *IEDM Tech. Digest*, 739 (1995).

[18] M.J.O. Strutt, *Semiconductor Devices*, Vol. 1, Academic Press, London (1966).

[19] S.M. Sze and G. Gibbons, 'Avalanche breakdown voltages of abrupt and linearly graded *pn* junctions in Ge, Si, GaAs, and GaP', *App. Phys. Lett.*, 8, 111 (1966).

[20] S.L. Miller, 'Ionization rates for electrons and holes in silicon', *Phys. Rev.* 99, 1234 (1955).

5

High-frequency Performance

5.1 INTRODUCTION

The high-frequency performance of bipolar transistors is determined by the minority carrier charge stored in the different regions of the transistor. This charge has to be removed from the transistor before it can turn off and hence it determines the maximum frequency at which the transistor is capable of operating. A transit time can be defined that is given by the ratio of the stored charge and the collector current. In forward active operation, this transit time is known as the forward transit time τ_F and represents a fundamental limit for the switching speed and maximum frequency of operation of a bipolar transistor. In this chapter, an expression will be derived for the forward transit time in terms of the physical parameters of the transistor.

In analogue circuits, the maximum frequency of operation of a bipolar transistor is of vital interest. The parameter most commonly used to define this frequency is the cut-off frequency f_T of the bipolar transistor. This is the frequency at which the common emitter, small-signal current gain drops to zero under conditions of a short-circuit load. An expression for the cut-off frequency will be derived and related to the forward transit time.

In practice parasitic capacitance and resistance will slow down the switching of digital bipolar circuits and limit the frequency of operation of analogue bipolar circuits. The origins of parasitic resistances and capacitances will be described, along with their effect on the high-frequency transistor performance. It will be shown that the

SiGe Heterojunction Bipolar Transistors Peter Ashburn
© 2003 John Wiley & Sons, Ltd ISBN: 0-470-84838-3

maximum oscillation frequency f_{max} is a good predictor of transistor performance, since it includes base resistance and collector/base capacitance, two of the most important parasitics associated with a bipolar transistor.

5.2 FORWARD TRANSIT TIME τ_F

The forward transit time models the excess charge stored in the transistor when its emitter/base junction is forward biased and its collector/base junction zero biased. This is an extremely important parameter, since it provides a fundamental physical limit to the switching speed and maximum frequency of operation of a bipolar transistor. In this section we will therefore consider this parameter in more detail, beginning with a study of the components of τ_F and moving on to derive its relationship to the cut-off frequency f_T.

5.2.1 Components of τ_F

The forward transit time τ_F can be written as the sum of the individual delay times in the various regions of the transistor:

$$\tau_F = \tau_E + \tau_{EBD} + \tau_B + \tau_{CBD} \tag{5.1}$$

where τ_E, τ_{EBD}, τ_B and τ_{CBD} are associated with the excess minority carrier charge in the neutral emitter, the emitter/base depletion region, the base and the collector/base depletion region respectively. The emitter delay τ_E and the emitter/base depletion region delay τ_{EBD} are generally small compared with the other terms in equation (5.1), although in high-speed bipolar transistors they can contribute significantly to the total forward transit time [1]. τ_B is associated with the excess minority carrier charge in the base and is referred to as the base delay, or more frequently the base transit time. τ_{CBD} is the collector/base depletion layer delay, and in high-speed bipolar transistors it is often of a similar magnitude to the base transit time.

5.2.2 Base Transit Time

The base transit time τ_B can be calculated from the minority carrier profile in the base, as shown in Figure 5.1. The charge in the base Q_b is

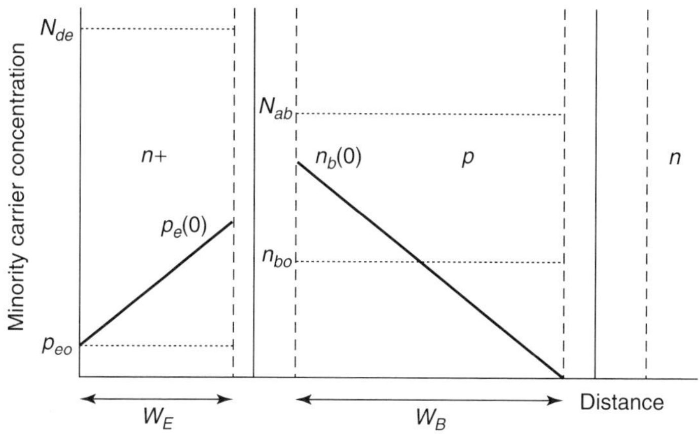

Figure 5.1 Minority carrier distributions in the base and emitter of a bipolar transistor

related to the area of the triangle created by the electron distribution in the base:

$$\begin{aligned} Q_b &= qA \text{ (area of triangle)} \\ &= qA\frac{1}{2}\,W_B n_b(0) = qA\frac{1}{2}\,W_B n_{bo} \exp\frac{qV_{BE}}{kT} \end{aligned} \tag{5.2}$$

where equation (2.30) has been used for $n_b(0)$. The base transit time is defined as the ratio of the charge stored in the base to the collector current:

$$\tau_B = \frac{Q_b}{I_C} \tag{5.3}$$

Using equation (2.40) for I_C gives:

$$\tau_B = \frac{W_B^2}{2D_{nb}} \tag{5.4}$$

Equation (5.4) shows that the base transit time is proportional to the square of the basewidth. In the design of high-speed bipolar transistors there is therefore a strong incentive to produce transistors with as small a basewidth as is practical.

Equation (5.4) is valid for bipolar transistors with a uniformly doped base. If the base is non-uniformly doped, as in the case of a bipolar transistor with an ion implanted base, then the variation in doping gives rise to a built-in electric field across the neutral base region. This is illustrated in Figure 5.2, which shows the band diagram for a

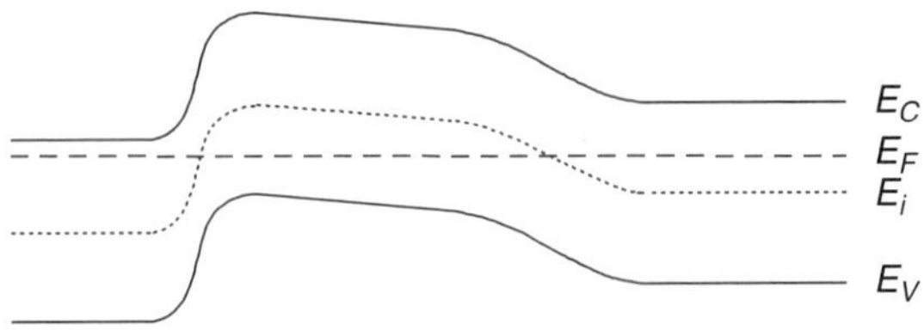

Figure 5.2 Band diagram of a bipolar transistor, illustrating how a decrease in base doping from emitter to collector gives rise to an accelerating built-in electric field

bipolar transistor in which the base doping decreases on moving from the emitter side of the base to the collector side. This gives rise to a downward slope on the conduction and valence bands on moving from emitter to collector. Minority carrier electrons in the base would see this slope as a built-in accelerating field that would aid electron transport across the base and hence reduce the base transit time. This situation can be taken into account by writing the equation for the base transit time as:

$$\tau_B = \frac{W_B^2}{\eta D_{nb}} \tag{5.5}$$

where η is a constant that has a value between 2 and 4 when an accelerating built-in field is present. Of course it is also possible to have a retarding field in the base if the base doping increases on moving from the emitter to collector. In this case, η would take a value between 1 and 2. When designing ion implanted bipolar transistors it is clearly important to maximize the accelerating built-in field.

5.2.3 Emitter Delay

The charge in the emitter Q_e can be calculated from the area of the triangle created by the minority carrier hole distribution in the emitter, as shown in Figure 5.1:

$$\begin{aligned} Q_e &= qA \text{ (area of triangle)} \\ &\approx qA\frac{1}{2}W_E p_e(0) = qA\frac{1}{2}W_E p_{eo} \exp\frac{qV_{BE}}{kT} \end{aligned} \tag{5.6}$$

The emitter delay τ_E is defined as the ratio of the charge in the emitter to the collector current:

$$\tau_E = \frac{Q_e}{I_C} \tag{5.7}$$

Using equation (3.19) for I_C gives:

$$\tau_E = \frac{W_E}{2N_{deff}} \frac{W_B N_{aeff}}{D_{nb}} \tag{5.8}$$

where the effects of doping induced bandgap narrowing have been included. When designing high-speed bipolar transistors, this equation shows that the emitter depth W_E should be as small as is practically possible and the emitter doping as high as possible.

5.2.4 Collector/Base Depletion Region Transit Time

The collector/base depletion layer delay τ_{CBD} is determined by the time required for electrons to traverse the base/collector depletion region. Electrons travel across the collector/base depletion region by drift, and hence this current can be written as:

$$I_n = qA\mu_n nE = qAnv_n \tag{5.9}$$

where n is the electron concentration in the collector/base depletion region and v_n is the drift velocity. The electric field across the collector/base depletion region is very high and hence the electrons reach their saturated velocity relatively quickly. Equation (5.9) can then be written as:

$$I_n = qAnv_{sat} \tag{5.10}$$

where the saturation velocity v_{sat} has a value of $1 \times 10^7\,\text{cms}^{-1}$ at room temperature.

At first sight, it would be expected that τ_{CBD} would be given by W_{CBD}/v_{sat}, where W_{CBD} is the collector/base depletion width. However, the situation is a little more complicated than this because the electron concentration in the depletion region changes when the depletion width changes. A rigorous analysis [2] shows that τ_{CBD} is given by:

$$\tau_{CBD} = \frac{W_{CBD}}{2v_{sat}} \tag{5.11}$$

From equation (5.11) it can be seen that the collector/base depletion layer delay can be reduced by decreasing the width of the collector/base depletion region. This can be achieved by increasing the doping concentration in the collector.

5.2.5 Emitter/Base Depletion Region Delay

The emitter/base depletion region delay is very small and to a first order can be neglected when calculating the forward transit time. For the purposes of compact transistor modelling where more accuracy is required, the small delay associated with the emitter/base depletion region can be modelled using a second-order correction to the emitter/base depletion capacitance or to the base transit time [3].

5.3 CUT-OFF FREQUENCY f_T

The most important high-frequency parameter for a bipolar transistor is the frequency at which the gain of the bipolar transistor drops to unity, otherwise known as the cut-off frequency f_T, as illustrated in Figure 5.3. Beyond this frequency the gain of the transistor is less than unity, so it is no longer useful as either an amplifying or a switching device. In practice, it becomes increasingly difficult to design circuits as the required circuit operating frequency approaches the cut-off frequency of the transistor. The rule of thumb used by many circuit designers is that circuit operation up to a factor of about ten below the cut-off frequency can be expected. Thus a bipolar transistor with an f_T of 50 GHz is suitable for use in circuits that operate up to around 5 GHz. Circuit operation at frequencies closer to the cut-off frequency of the bipolar transistor can be achieved with careful circuit design. If you are reading this book for the first time, or merely interested in the final result, you may wish to skip the following derivation and move on to equation (5.19).

The cut-off frequency of a bipolar transistor is defined as the frequency at which the extrapolated common emitter, small-signal current gain

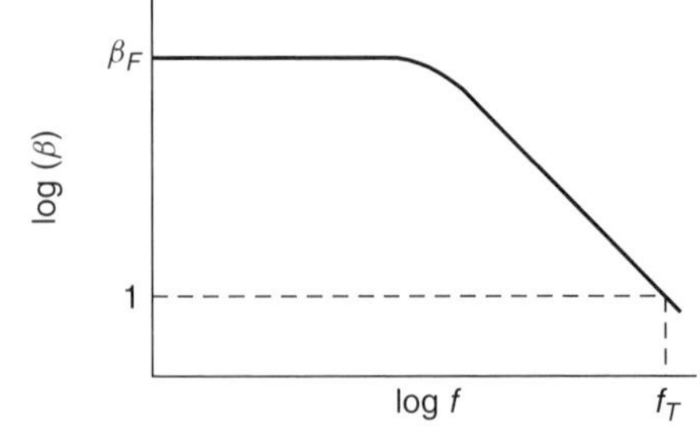

Figure 5.3 Variation of gain with frequency and the definition of the cut-off frequency f_T

drops to unity under conditions of a short-circuit load. Since f_T is defined for small-signal conditions, a small-signal circuit model, such as the hybrid-π model, can be used to derive an expression for f_T. Circuit models of bipolar transistors will be discussed in Chapter 11, but for the time being, the hybrid-π model is illustrated in Figure 5.4, with a short-circuit applied to the load. From this circuit, the collector and base currents can be written as:

$$i_c = g_m v_{be} - j\omega C_\mu v_{be} \tag{5.12}$$

$$i_b = v_{be}(g_\pi + j\omega C_\pi + j\omega C_\mu) \tag{5.13}$$

where $g_\pi = 1/r_\pi$ and $\omega = 2\pi f$. The common emitter current gain can therefore be written as:

$$\beta = \frac{i_c}{i_b} = \frac{g_m - j\omega C_\mu}{g_\pi + j\omega C_\pi + j\omega C_\mu} \tag{5.14}$$

At most frequencies of practical interest, $g_m \gg j\omega C_\mu$ and hence equation (5.14) can be simplified to:

$$\beta = \frac{\beta_F}{1 + j\omega r_\pi (C_\pi + C_\mu)} \tag{5.15}$$

where we have used equations (11.23) and (11.25) to define β_F as:

$$\beta_F = g_m r_\pi \tag{5.16}$$

It can be seen from equation (5.15) that the common emitter current gain has a value of β_F at low frequencies, as illustrated in Figure 5.3. At

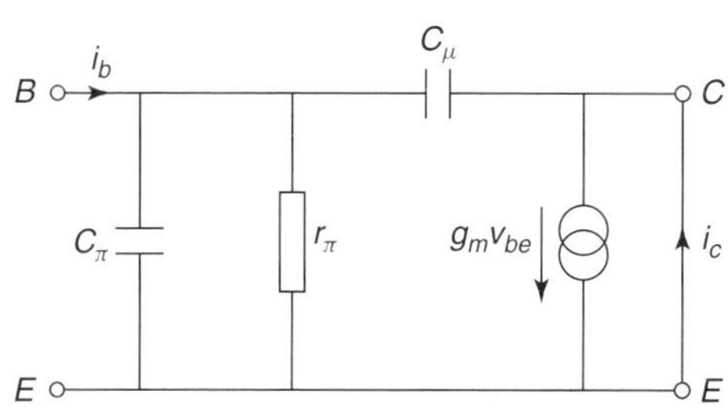

Figure 5.4 Use of the small-signal hybrid-π model for the calculation of the cut-off frequency

high frequencies the second term in the denominator of equation (5.15) is large with respect to unity, and β can be approximated by:

$$|\beta| = \frac{\beta_F}{\omega r_\pi (C_\pi + C_\mu)} \tag{5.17}$$

The common emitter current gain falls to unity when:

$$1 = \frac{\beta_F}{2\pi f r_\pi (C_\pi + C_\mu)} \tag{5.18}$$

Rearranging and using equations (11.23) and (11.28) gives the following expression for the cut-off frequency:

$$f_T = \frac{1}{2\pi \left(\tau_F + \dfrac{kT}{qI_C}(C_{JE} + C_{JC}) \right)} \tag{5.19}$$

where C_{JE} and C_{JC} are the emitter/base and base/collector depletion capacitances and τ_F is given by equation (5.1).

For completeness, an additional term should be included in equation (5.19) to account for the RC delay due to the series collector resistance and the collector/base capacitance [2,6]. The complete equation for the cut-off frequency then becomes:

$$f_T = \frac{1}{2\pi \left(\tau_F + R_C C_{JC} + \dfrac{kT}{qI_C}(C_{JE} + C_{JC}) \right)} \tag{5.20}$$

The dependence of f_T on collector current is illustrated in Figure 5.5. At low currents the depletion capacitance term in equation (5.20) is much larger than the other two terms, and hence f_T increases with I_C. At medium currents the depletion capacitance term becomes smaller than τ_F, and hence f_T ceases to rise with collector current. In this part of the characteristic f_T is equal to f_{TMAX}, and is given by:

$$f_{TMAX} = \frac{1}{2\pi \tau_F} \tag{5.21}$$

At high collector currents the cut-off frequency decreases markedly due to high current effects, which will be described in Section 5.5. In many

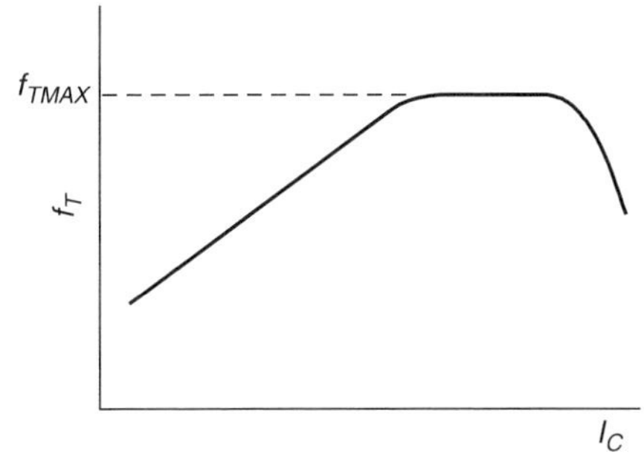

Figure 5.5 Variation of the cut-off frequency f_T with collector current

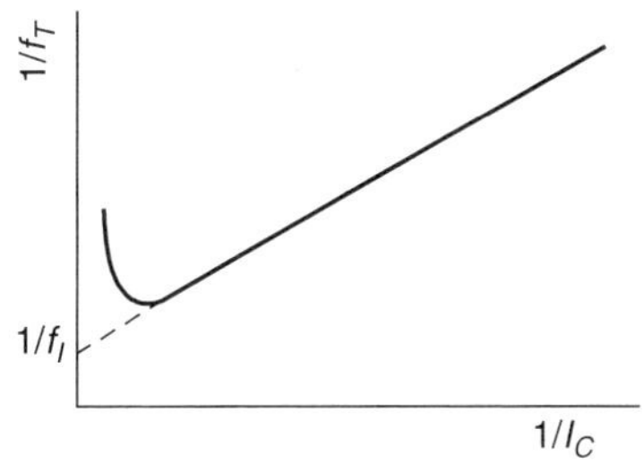

Figure 5.6 Method of measuring the τ_F of a bipolar transistor

transistors a clearly defined region of constant-f_T is not discernable. However, in this case the forward transit time can be obtained from a graph of l/f_T versus l/I_C, as illustrated in Figure 5.6. The intercept of the extrapolated straight line with the vertical axis can be used to calculate τ_F:

$$\tau_F = \frac{1}{2\pi f_I} - R_C C_{JC} \tag{5.22}$$

5.4 MAXIMUM OSCILLATION FREQUENCY f_{max}

Another important high-frequency parameter for a bipolar transistor is the maximum oscillation frequency f_{max}. This is defined as the frequency at which the power gain of a bipolar transistor drops to unity. An approach similar to that in Section 5.3 [2] can be used to derive an expression for the f_{max} of a bipolar transistor:

$$f_{max} = \sqrt{\frac{f_T}{8\pi C_{JC} R_B}} \tag{5.23}$$

This equation shows that the f_{max} of a bipolar transistor is determined not only by the f_T but also by collector/base capacitance C_{JC} and the base resistance R_B. These two parasitics have a strong influence on the performance of bipolar circuits, and hence the f_{max} is a better predictor of circuit performance than the f_T. Bipolar transistor design requires a compromise between f_T, C_{JC} and R_B.

5.5 KIRK EFFECT

The analysis in Section 5.2 suggests that the forward transit time should be constant and independent of current. Although this is true at low currents, at high currents τ_F increases markedly with collector current [4,5]. The reason for this is an increase in the effective basewidth of the transistor [4] at high currents due to a current-dependent build-up of the minority carrier charge in the collector/base depletion region. This occurs when the mobile charge in the collector/base depletion region becomes greater than the fixed ionized charge, and this leads to the spreading of the neutral base region into the collector at high current densities. This effect is known as base widening or the Kirk effect [4]. Two-dimensional spreading effects [5] can also contribute to this degradation of τ_F at high collector currents.

The base-widening effect can be best understood by considering Poisson's equation (2.13) as applied to the collector/base depletion region:

$$\frac{dE}{dx} = \frac{\rho}{\varepsilon_0\varepsilon_r} = \frac{q}{\varepsilon_0\varepsilon_r}(p - n + N_{dc}) \tag{5.24}$$

where p and n represent the mobile charge and N_{dc} the fixed ionized charge in the depletion region. In an *npn* transistor the current is carried predominantly by electrons, and hence $p \approx 0$ in equation (5.24). As discussed above, in the collector/base depletion region the electrons are transported by drift, and hence the electron concentration in the collector/base depletion region can be obtained from equation (5.10). Combining equations (5.24) and (5.10) gives:

$$\frac{dE}{dx} = \frac{1}{\varepsilon_0\varepsilon_r}\left(qN_{dc} - \frac{J_n}{v_{sat}}\right) \tag{5.25}$$

In the simple theory in Section 5.2, it was implicitly assumed that the mobile charge in the depletion region J_n/v_{sat} was much smaller than the fixed charge qN_{dc}. However, it can be seen from equation (5.25) that this is only true if $J_n \ll qN_{dc}v_{sat}$. For a collector doping concentration of $1 \times 10^{16}\,\text{cm}^{-3}$ this is equivalent to an electron current density of $1.6 \times 10^4\,\text{Acm}^{-2}$. In practice, many bipolar transistors operate at or above this current, and hence it is clear that base widening effects are very important.

Base widening can be understood physically by considering Figure 5.7. At low current densities, charge density and field distributions are as shown in Figure 5.7(a). The mobile electron concentration in the collector/base depletion region is small with respect to the doping concentration in the collector epitaxial layer N_{dc} and hence it has no effect on the depletion width. The fixed positive donor charge in the *n*-type collector is balanced by the fixed negative acceptor charge in the base and the electric field is a maximum at the metallurgical collector/base junction.

As the collector current density increases, the mobile electron concentration in the collector/base depletion region increases. Eventually a point is reached where the mobile electron concentration is high enough to begin compensating the fixed positive donor charge. Since the charge at either side of the collector/base depletion region must exactly balance, the depletion width in the collector must increase to bring the net positive charge into balance with the fixed negative acceptor charge. This situation is illustrated in Figure 5.7(b), where the mobile electron concentration in the collector/base depletion region is assumed to be half of the collector doping concentration N_{dc}. Consequently the depletion region width needs to double to bring the net positive charge in the collector side of the depletion region into balance with the fixed negative acceptor charge at the base side.

As the collector current density further increases, the collector/base depletion region extends deeper into the collector until it reaches the buried layer, as illustrated in Figure 5.7(c). At this point the epitaxial collector is completely depleted, and the electric field is constant across the epitaxial layer. The depletion region extends into the buried layer and the fixed positive donor charge in the buried layer balances the fixed negative acceptor charge in the base.

When the collector current density increases above the point shown in Figure 5.7(c), the mobile electron concentration in the collector/base depletion region becomes greater than the fixed positive donor charge. At this point, the gradient of the electric field in the epitaxial collector reverses, as shown in Figure 5.7(d). The field distribution can best be understood by starting in the buried layer, where the field is zero. As we approach the epitaxial collector, the field rises from zero to reach a

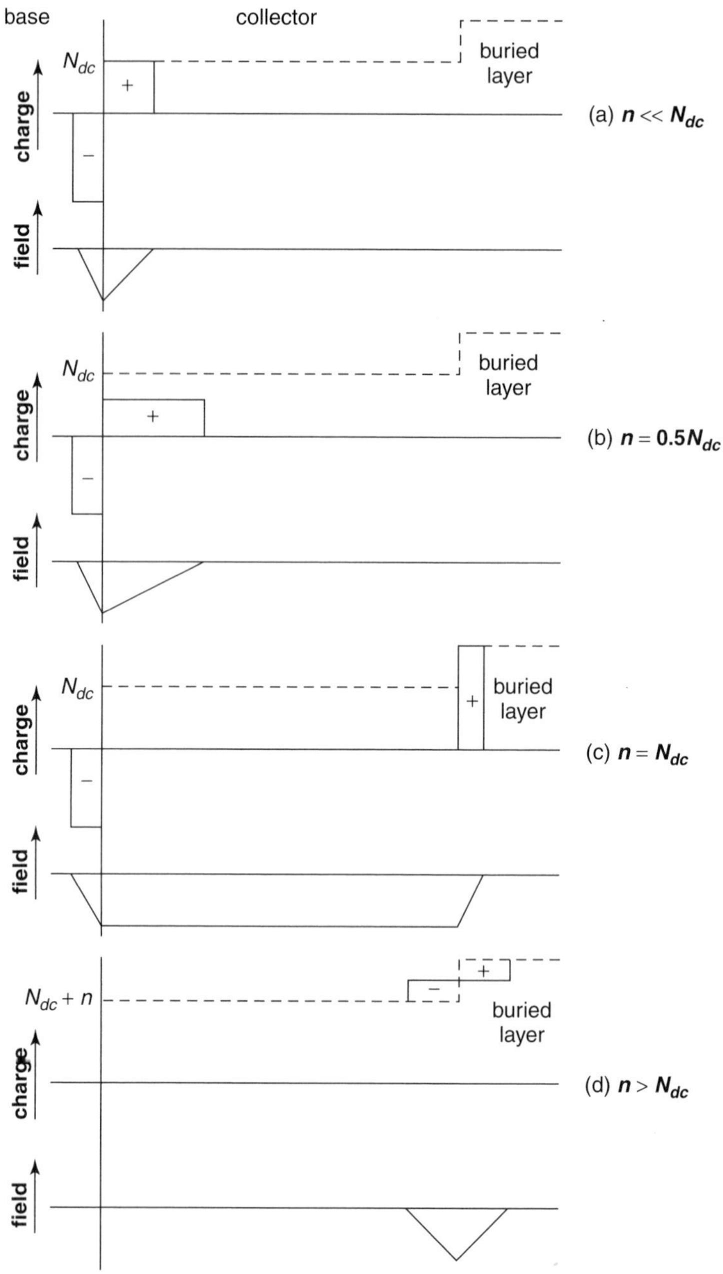

Figure 5.7 Charge and field distributions in the collector/base depletion region at different current densities

maximum just inside the buried layer. Moving beyond this point, the only possible solution to the equations that gives a reversed electric field gradient is for the electric field to decrease to zero somewhere in the epitaxial collector. This is illustrated in Figure 5.7(d), and from the field distribution it can be seen that the neutral base has widened and pushed into the epitaxial collector. This widening occurs by the accumulation of mobile holes in the epitaxial collector.

It is clear from the complex set of field and charge distributions in Figure 5.7 that the quantification of the Kirk effect requires numerical simulation. However, the effects of the Kirk effect on the transistor performance can be qualitatively inferred from Figure 5.7. The widening of the collector/base depletion region seen in Figure 5.7(b) will lead to an increase in the collector/base depletion region delay as can be seen from equation (5.11). This in turn will lead to an increase in τ_F (equation (5.1)) and a decrease in f_T. The spreading of the base into the collector shown in Figure 5.7(d) will lead to an increase in base transit time (equation (5.5)), which again will lead to an increase in τ_F and a decrease in f_T. Furthermore, the spreading of the base into the epitaxial collector will increase the effective basewidth and hence cause a decrease in the gain, as can be seen from equation (3.20).

The current at which the onset of the Kirk effect occurs can be determined from equation (5.25), and occurs when the two terms in brackets are comparable in magnitude. The two terms are equal at a current density of:

$$J_n = qN_{dc}v_{sat} \tag{5.26}$$

For a collector doping of $1 \times 10^{16}\,\text{cm}^{-3}$ this is equivalent to a current density of $16\,\text{kA/cm}^2$

In order to give a better understanding of the behaviour of a bipolar transistor at high currents, Table 5.1 gives a breakdown of the components of f_T at different collector current densities. The figures shown are for a silicon high-speed bipolar transistor, and the results were computed using the BIPOLE device simulation program [7]. The delay τ_{RE} represents the depletion capacitance term in equation (5.20), and the other delays are the components of τ_F. At low collector currents it can be seen that τ_{RE} is by far the dominant component of f_T, as expected from equation (5.20). However, at collector currents around the peak f_T all the terms contribute significantly to the total delay, though τ_{RE}, τ_{CBD} and τ_B are the largest. The decrease in f_T at high collector currents

Table 5.1 Breakdown of components of f_T at different values of collector current density

Collector current density, A/cm^2	Delay components of f_T, ps					f_T, GHz	Effective basewidth, μm
	τ_E	τ_{EBD}	τ_B	τ_{CBD}	τ_{RE}		
2.9×10^2	0.23	2.2	0.53	3.0	45	3.1	0.043
1.1×10^3	0.25	1.4	0.59	3.0	14	8.3	0.043
3.6×10^3	0.32	1.0	0.86	3.0	5.9	14.4	0.043
1.3×10^4	0.43	0.70	1.40	3.0	3.1	18.4	0.049
2.4×10^4	0.58	0.53	1.80	4.4	2.7	15.9	0.057

is due to the Kirk effect, as can be seen from the increase in the values of τ_{CBD} and τ_B at high currents. Base spreading into the epitaxial collector can be inferred from the increase in effective basewidth at high currents.

5.6 BASE, COLLECTOR AND EMITTER RESISTANCE

As discussed in Section 4.5, the series resistance of the silicon in the base, collector and emitter gives rise to base, collector and emitter resistance. These series resistances influence the high-frequency performance of the bipolar transistor when they combine with parasitic capacitances to give RC time constants. Equation (5.20) shows that the collector resistance in combination with the collector/base capacitance influences the value of f_T and equation (5.23) shows that the base resistance in combination with the collector/base capacitance influences the value of f_{max}. Similarly, these series resistances influence the performance of bipolar circuits, as will be seen in Chapter 12. The emitter resistance is generally very small because the contact to the emitter is made directly above the emitter. The one exception to this statement is polysilicon emitters, where additional emitter resistance arises from the polysilicon/silicon interface, as will be seen in Chapter 6.

5.6.1 Base Resistance

Base resistance is one of the most important electrical parameters of a bipolar transistor. It limits the rate at which the input capacitance can be charged and is therefore one reason why bipolar transistors do not operate at the frequencies predicted by the values of forward transit time. The base resistance can be partitioned into two parts, the intrinsic and extrinsic base resistances, as illustrated in Figure 5.8. The total base

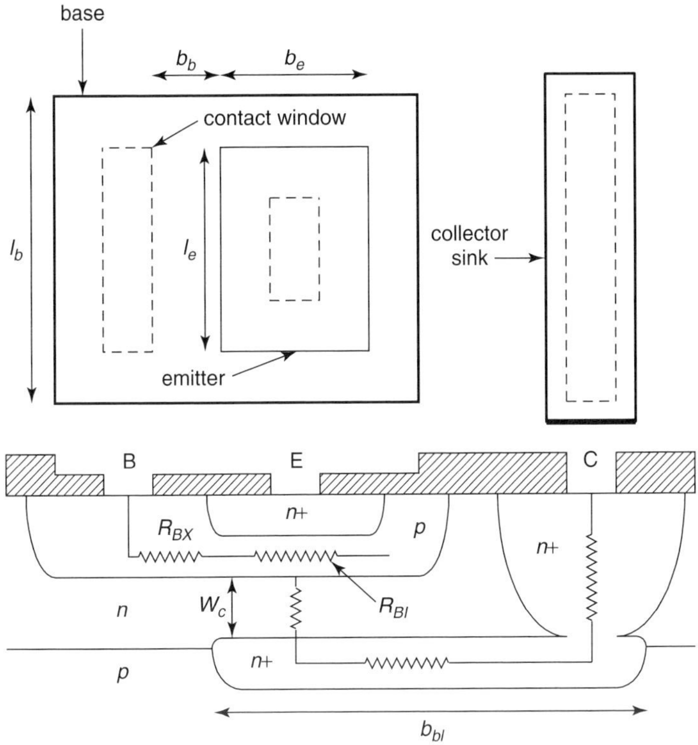

Figure 5.8 Schematic plan and cross-section views of a basic bipolar transistor. The plan view shows collector sink, base, emitter and contact window masks. The buried layer mask is not shown. The cross-section view shows the components of base and collector resistance

resistance is then given by the sum of these two components. Similarly the collector resistance has three components due to the resistances of the epitaxial collector, buried layer and collector sink.

The extrinsic base resistance R_{BX} is the resistance between the edge of the active transistor area and the base contact, and can be estimated from the transistor geometry and the extrinsic base sheet resistance R_{SBX}:

$$R_{BX} = \frac{R_{SBX}\dfrac{b_b}{l_b} + R_{CON}}{n_B} \tag{5.27}$$

where R_{CON} is the contact resistance and n_B is the number of base contacts. The intrinsic base resistance is the resistance of the active base region, which is the region located beneath the emitter. It can be estimated from the transistor geometry and the intrinsic base sheet resistance:

$$R_{BI} = C\frac{R_{SBI}\dfrac{b_e}{l_e}}{n_b^2} \tag{5.28}$$

where C is a constant that takes a value of 1/3 at low currents [2]. The explanation for the n_B^2 term in equation (5.28) is as follows. If the transistor has only one base contact, the base current enters from only one side of the emitter and hence the path length for the current flow is the complete emitter width. If the transistor has two base contacts, the base current enters from both sides of the emitter, so the path length for the current flow is halved. A further halving of the intrinsic base resistance arises because the two base contacts are in parallel. Equations (5.27) and (5.28) both show that the base resistance is reduced if two base contacts are used. However, this benefit is obtained at the expense of increased collector/base capacitance, because extra area is needed for the second base contact.

5.6.2 Collector Resistance

Collector resistance arises because of the planar structure of the bipolar transistor, as shown in Figure 5.8. Contact to the collector is made to one side of the base and the collector current must flow through the collector and up to the surface to reach the collector contact. Since the collector is relatively lowly doped, the collector resistance could be very large. To address this problem, a heavily doped buried layer is included below the base, which gives a low resistance path in parallel with the high resistance of the epitaxial collector. Furthermore, a heavily doped collector sink region is included below the collector contact to reduce the resistance from the buried layer to the collector contact. If these two features are included in the bipolar transistor fabrication process, the collector resistance is dramatically reduced and limited primarily by the resistance of the epitaxial collector beneath the base. Clearly in this situation, reducing the thickness of the epitaxial layer will lead directly to a reduction in collector resistance. Further information on the fabrication of the collector is given in Chapter 9.

The collector resistance can be estimated from the transistor geometry and the sheet resistances of the epitaxial collector R_{SC} and the buried layer R_{SBL}:

$$R_C = R_{SBL}\frac{b_{bl}}{l_{bl}} + R_{SC}\frac{W_C^2}{b_c l_c} + R_{CC} \tag{5.29}$$

where l_{bl} (b_{bl}) is the length (width) of the buried layer, l_c (b_c) the length (width) of the collector region, and R_{CC} is the collector contact resistance, which includes a small contribution from the collector sink. Other parameters are shown in Figure 5.8. The parameters l_c and b_c are determined by the bounds of the isolation region, which is not shown in Figure 5.8. As will be discussed in Chapter 9, in many bipolar technologies oxide isolation is used to exactly align the isolation to the base. In this case, l_c would be the same as l_b and b_c the same as b_b.

5.7 EMITTER/BASE AND COLLECTOR/BASE DEPLETION CAPACITANCE

The fixed charges in the depletion regions of the emitter/base and collector/base junctions give rise to capacitances, denoted by the emitter/base junction capacitance C_{JE} and the collector/base junction capacitance C_{JC}. Collector/base capacitance is often partitioned into intrinsic and extrinsic components in the same way as base resistance, as illustrated in Figure 5.9. The intrinsic collector/base capacitance is determined by the emitter size and the extrinsic collector/base capacitance is determined by the space required to make a contact to the base. It will be shown in Chapter 9 that special fabrication techniques can be used to dramatically reduce the extrinsic collector/base capacitance.

The emitter/base and collector/base depletion capacitances are given by the standard textbook expressions for depletion capacitance. For example the emitter/base depletion capacitance is given by:

$$C_{JE} = \frac{C_{JE0}}{\left(1 - \frac{V_{BE}}{V_{JE}}\right)^{M_{JE}}} \tag{5.30}$$

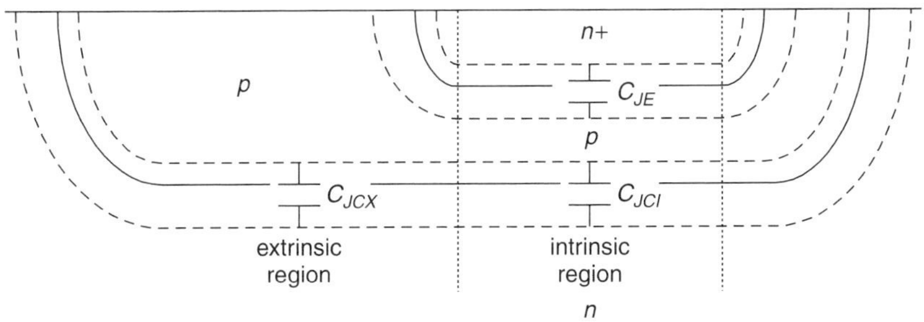

Figure 5.9 Schematic illustration of emitter/base and collector/base depletion capacitances in a bipolar transistor

where C_{JE0} is the value of emitter/base capacitance at zero bias, V_{JE} is the junction built-in voltage and M_{JE} is a factor that is determined by the gradient of the emitter profile. For an abrupt profile M_{JE} has a value of 1/2, whereas for a linearly graded profile it has a value of 1/3. A similar equation is used for the collector/base depletion capacitance.

$$C_{JC} = \frac{C_{JC0}}{\left(1 - \dfrac{V_{BC}}{V_{JC}}\right)^{M_{JC}}} \tag{5.31}$$

5.8 QUASI-SATURATION

Quasi-saturation is an effect that occurs at high currents due to the internal collector resistance of the bipolar transistor. It occurs when the voltage drop across the collector resistance is large enough to forward bias the collector/base junction.

Figure 5.10 shows the circuit diagram of a bipolar transistor including the internal collector resistance R_C. The internal collector/base voltage $V_{C'B'}$ is given by:

$$V_{C'B'} = V_{CB} - I_C R_C \tag{5.32}$$

If the external collector/base voltage V_{CB} is fixed and the collector current increased, equation (5.32) shows that the internal collector/base voltage $V_{C'B'}$ will become negative at high values of collector current. In other words, the collector/base junction will become forward biased, injecting charge into both the base and the collector. When this forward bias exceeds about 0.5 V, the injected charge will have the effect of decreasing the current gain.

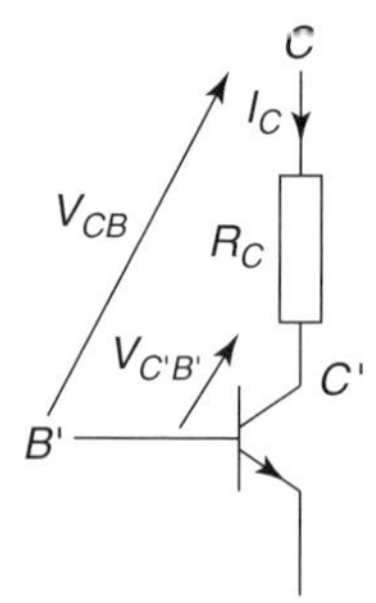

Figure 5.10 Circuit diagram showing the internal collector resistance due to the series resistance of the epitaxial layer and buried layer

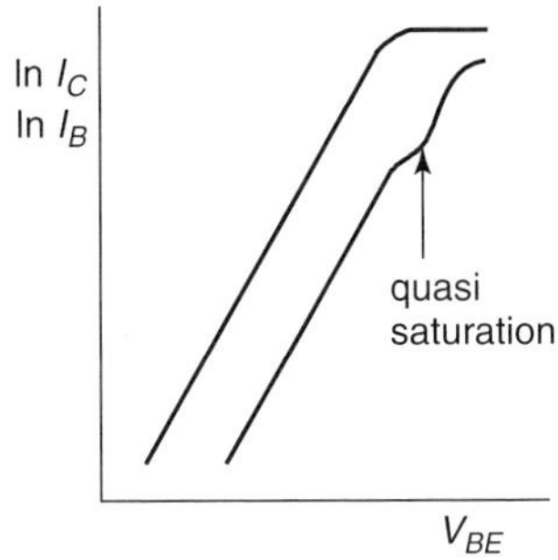

Figure 5.11 Gummel plot showing the effect of quasi-saturation

Quasi-saturation is most likely to be seen in configurations where the external collector/base voltage is small. This is the situation when Gummel plots are measured, since the external collector/base bias is maintained at 0 V. Quasi-saturation is therefore often seen in Gummel plots at high currents, as illustrated in Figure 5.11. The kink in the base characteristics indicates the onset of quasi-saturation, and beyond this point the gain decreases. A simple test for quasi-saturation is to increase the collector/base reverse bias. If quasi-saturation is responsible for the kink in the base characteristic, then an increase in the collector/base reverse bias should cause the kink to move to higher base/emitter voltages.

Quasi-saturation can also be seen in the output characteristics of bipolar transistors at high currents, as illustrated in Figure 5.12. The quasi-saturation region can be seen at low values of collector/emitter voltage, where the forward biasing of the collector/base junction can occur. The quasi-saturation is seen as a soft transition into the saturation

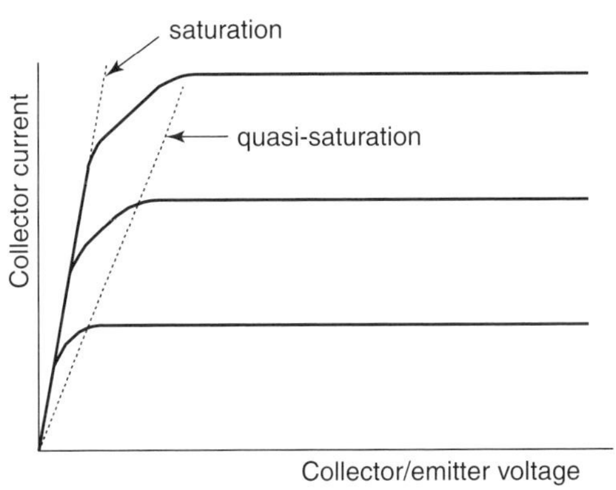

Figure 5.12 Output characteristic showing the effect of quasi-saturation

region of the output characteristic. Its effect is greater at higher collector currents, because the voltage drop across the internal collector resistor is larger.

5.9 CURRENT CROWDING

At high currents there can be a significant voltage drop across the intrinsic base resistance of a bipolar transistor due to the lateral flow of base current from the base contact. This situation is illustrated in Figure 5.13. As a result of this voltage drop across the intrinsic base resistance, the potential in the base close to the base contact is higher than that away from the base contact. Consequently the emitter/base junction is more forward biased at the edge of the emitter that is closest to the base contact. Equations (3.18) and (3.19) show that both the base current and collector current will be higher at this point. This effect is known as current crowding since the current crowds to the edge of the emitter that is closest to the base contact. In transistors with two base contacts, the current crowds to the perimeter of the emitter. If a bipolar transistor is required to deliver a high current, it is therefore necessary to maximize the emitter perimeter for a given emitter area. This can be done by partitioning the emitter into separate emitter fingers.

Current crowding has a strong effect on the base resistance of the bipolar transistor. At low currents, the voltage drop across the intrinsic base resistance is small, so the current is evenly distributed across the emitter. In this case the total base resistance is given by the sum of the extrinsic and intrinsic base resistances, as given in equations (5.27) and (5.28). At high currents, the current crowds towards the perimeter of the emitter, so the current does not see all of the intrinsic base resistance. The net result is that the total base resistance decreases with increasing current and approaches the value for the extrinsic base resistance. Current crowding

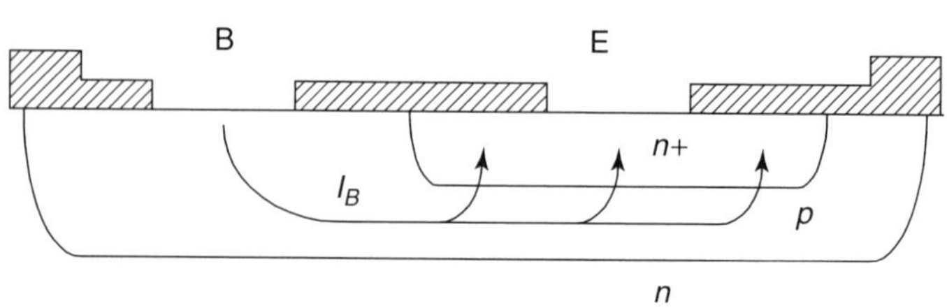

Figure 5.13 Schematic cross-section of a bipolar transistor showing the flow of hole current from the base contact

is beneficial for the performance of bipolar transistors at high frequency, since the decrease in base resistance reduces the delay associated with the base resistance time constants.

REFERENCES

[1] J.A. Kerr and F. Berz, 'The effect of emitter doping gradient on f_T in microwave bipolar transistors', *IEEE Trans. Electron. Devices*, **22**, 15 (1975).

[2] D.J. Roulston, 'Bipolar semiconductor devices', 239, McGraw Hill (1990).

[3] J.J.H. van den Biessen, 'A simple regional analysis of transit times in bipolar transistors', *Solid State Electronics*, **29**, 529 (1986).

[4] C.T. Kirk, 'A theory of transistor cut-off frequency falloff at high current densities', *IRE Trans. Electron. Devices*, **9**, 164 (1962).

[5] A. van der Ziel and D. Agouridis, 'The cut-off frequency fall-off in UHF transistors at high currents', *Proc. IEEE*, **54**, 411 (1966).

[6] I.E. Getreu, *Modelling the Bipolar Transistor*, Elsevier, Amsterdam (1978).

[7] D.J. Roulston, S.G. Chamberlain and J. Sehgal, 'Simplified computer aided analysis of double diffused transistors including two-dimensional high-level effects', *IEEE Trans. Electron. Devices*, **19**, 809 (1972).

6

Polysilicon Emitters

6.1 INTRODUCTION

Scaling of bipolar transistors, like the scaling of MOS transistors, requires that the junction depth be reduced when the minimum feature size is reduced [1–3]. The primary reason for this requirement is that the emitter/base capacitance does not scale properly unless the emitter/base junction depth is scaled in proportion. This can be understood from Table 6.1, which shows calculated values of emitter/base junction capacitance. The capacitance has been partitioned into a peripheral component due to the junction sidewalls and a plane component due to the bottom junction. For the purposes of this comparison the peripheral capacitances have been calculated assuming a cylindrical shape for the lateral diffusion. At an emitter geometry of $1.5 \times 1.5\,\mu m$ and a junction depth of $0.2\,\mu m$ the peripheral capacitance contributes nearly 50% of the total capacitance. If the emitter size is reduced to $0.5 \times 0.5\,\mu m$ and the emitter/base junction depth maintained at $0.2\,\mu m$, the plane capacitance scales by a factor of 9, but peripheral capacitance only scales by a factor of 3. The net result is that the peripheral capacitance contributes nearly 70% to the total emitter/base capacitance. To obtain the full benefits of scaling, the emitter/base junction depth must be scaled in the same proportion as the emitter size.

There are problems in scaling the emitter/base junction depth W_E of a bipolar transistor because the gain reduces when W_E is reduced, as can be seen from equation (3.20). This difficulty in scaling posed a serious problem for bipolar technology until the emergence of polysilicon emitters provided a solution. It will be shown that polysilicon

SiGe Heterojunction Bipolar Transistors Peter Ashburn
 ISBN: 0-470-84838-3

Table 6.1 Comparison of plane and peripheral components of emitter/base capacitance

Emitter size, μm	Emitter/base junction depth, μm	Plane capacitance, fF	Peripheral capacitance, fF
1.5 × 1.5	0.2	4.1	3.0
1.5 × 1.5	0.1	4.1	1.6
1.5 × 1.5	0.02	4.1	0.4
0.5 × 0.5	0.2	0.46	0.99
0.5 × 0.5	0.1	0.46	0.55
0.5 × 0.5	0.02	0.46	0.14

emitters allow the emitter/base junction depth to be scaled without a degradation of gain [4]. Another big advantage of polysilicon emitters is their compatibility with self-aligned fabrication techniques [5]. These techniques will be described in detail in Chapter 9, but in essence they allow the parasitic resistances and capacitances of a bipolar transistor to be minimized, with the result that a considerable improvement in circuit performance is obtained. In this chapter, the theory and practice of polysilicon emitters will be described in detail.

6.2 BASIC FABRICATION AND OPERATION OF POLYSILICON EMITTERS

Polysilicon emitters are formed using polycrystalline silicon, which is a form of silicon part-way between perfectly ordered single-crystal silicon and totally unordered amorphous silicon. It consists of small, randomly oriented grains of single-crystal silicon, separated by disordered regions known as grain boundaries, as illustrated schematically in Figure 6.1. Polysilicon is a technologically important material because it is widely used in MOS processes to form the gate electrode of the MOS transistor. It is particularly useful for polysilicon emitters because of its low deposition temperature (≈600°C) and its ability to withstand the high temperatures routinely used in the fabrication of integrated circuits (900–1050°C). It also has useful electrical properties, since in many respects it behaves in away similar to single-crystal silicon. Thus it can be doped to produce *n*- or *p*-type layers, and at high doping concentrations, reasonably low sheet resistances can be achieved (≈50 Ω/sq).

The use of polysilicon emitters to give shallow emitter/base junctions can be understood from the fabrication sequence illustrated in Figure 6.2. After the opening of an emitter window, an undoped polysilicon layer

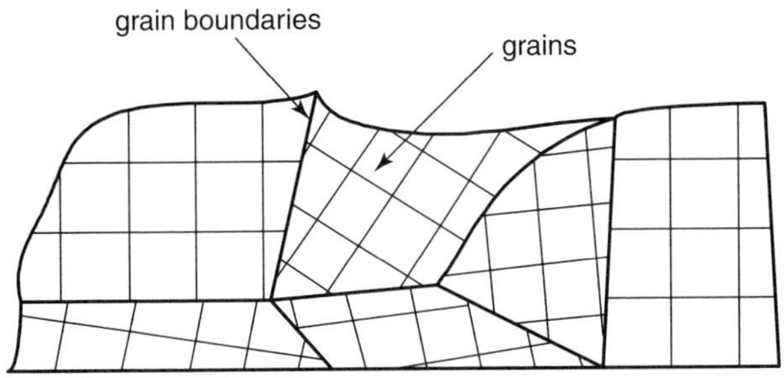

Figure 6.1 Schematic illustration of the structure of polysilicon

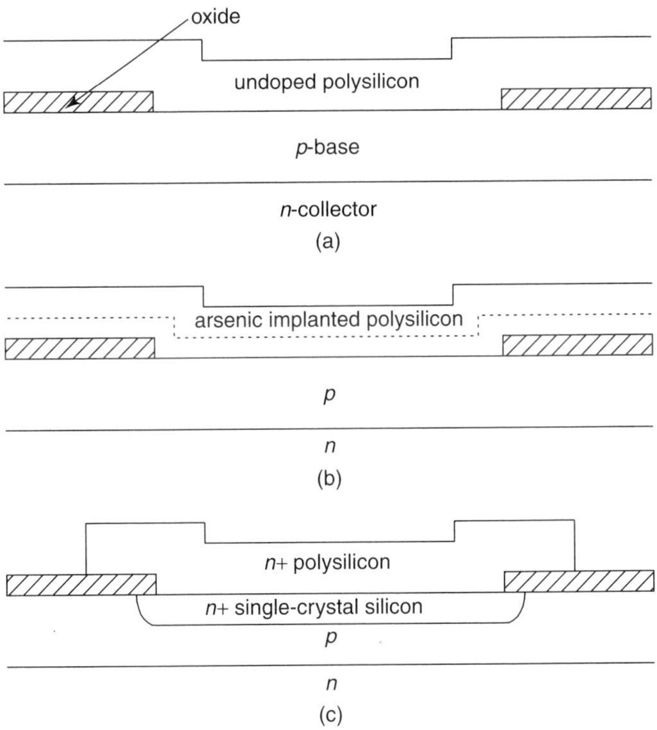

Figure 6.2 Fabrication sequence for a polysilicon emitter

is deposited onto the wafer, as shown in Figure 6.2(a). Arsenic is then implanted into the polysilicon at an energy that places the arsenic in the polysilicon layer, without giving any penetration into the underlying single-crystal silicon (Figure 6.2(b)). An anneal is carried out to diffuse the arsenic from the polysilicon into the underlying single-crystal silicon. The peripheral emitter/base capacitance is determined by the amount of arsenic penetration into the single-crystal silicon. This can be extremely shallow (down to $\approx$20 nm) because the heavily doped polysilicon layer acts as an ideal diffusion source. The polysilicon emitter is completed by defining and etching the polysilicon, as shown in Figure 6.2(c).

The factors influencing the gain of a polysilicon emitter can be understood from the minority carrier distribution. Figure 6.3(a) shows the situation for a conventional silicon emitter. It is assumed that the emitter is shallow, so the minority carrier hole distribution in the emitter is linear, as was shown in Figure 2.2. The base current is proportional to the gradient of the minority carrier distribution, as shown previously in equation (2.23). Figure 6.3(b) shows the situation when the emitter/base junction depth is scaled. It is clear that the slope of the minority carrier hole distribution is steeper than that in Figure 6.3(a), so the base current will be higher and the gain lower. Figure 6.3(c) shows the situation for a polysilicon emitter, where it has been assumed that the polysilicon behaves in the same way as single-crystal silicon. The minority carrier distribution is identical to that in Figure 6.3(a), even though the emitter/base junction depth is considerably shallower, so the base current and hence the gain will be the same. This qualitative analysis shows that the peripheral component of the emitter/base depletion capacitance is determined by the single-crystal emitter depth, while the base current (and gain) is determined by the sum of the single-crystal emitter depth and the polysilicon thickness. This allows low values of emitter/base capacitance to be achieved at the same time as high values of gain.

6.3 DIFFUSION IN POLYSILICON EMITTERS

The formation of the single-crystal part of the polysilicon emitter depends critically on the diffusion of the emitter dopant in the polysilicon. Arsenic is generally the preferred emitter dopant because it has been used for many years in both bipolar and MOS technologies and hence its behaviour is well understood. However, phosphorus can also be used [6], and has advantages where a low thermal budget is required because of its higher diffusion coefficient than arsenic. Both arsenic and phosphorus diffusion in polysilicon emitters occur in two distinct ways. The first is rapid diffusion down the grain boundaries, and the second is slower diffusion into the bulk of the grains and, at the same time, into the single-crystal emitter. Both arsenic and phosphorus segregate to grain boundaries [7], which has the advantage of giving a large concentration of dopant on the grain boundaries, which is then able to diffuse down the grain boundary to the polysilicon/silicon interface. The disadvantage is

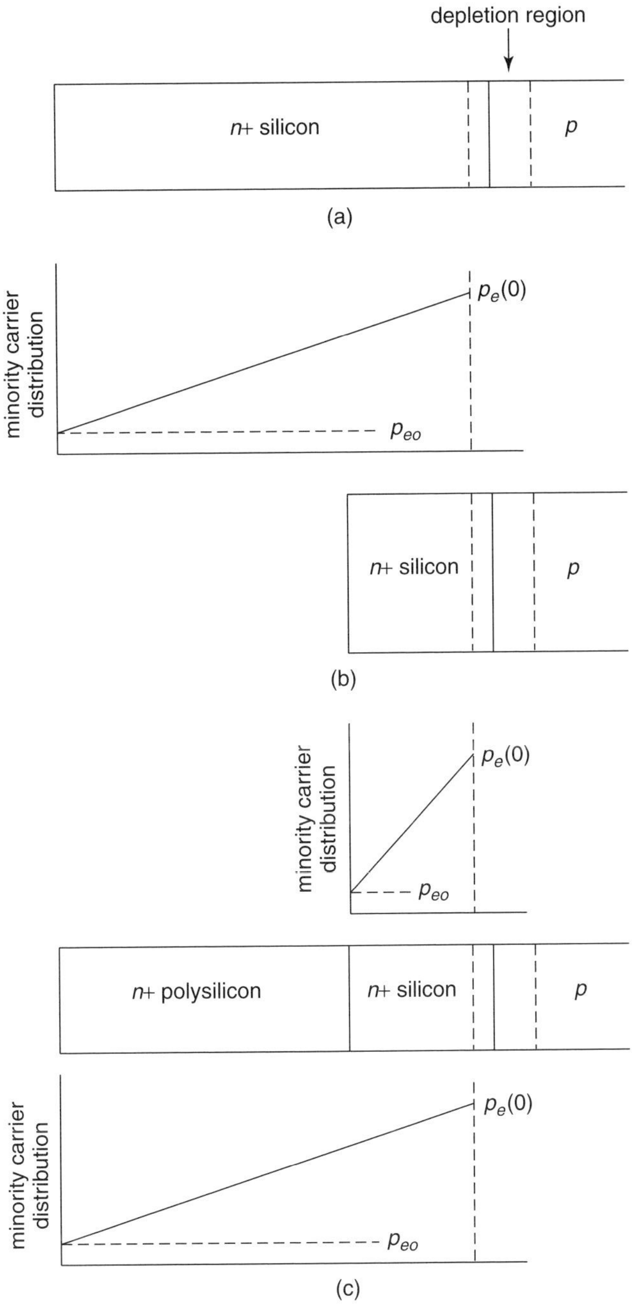

Figure 6.3 Minority carrier hole distributions in (a) a conventional silicon emitter, (b) a scaled, conventional silicon emitter and (c) a polysilicon emitter

that the grain boundaries have to be saturated with dopant to satisfy the segregation coefficient before significant dopant is available to diffuse into the grain interiors.

The mechanism of arsenic diffusion in polysilicon emitters is illustrated in Figure 6.4. After implant, the top part of the polysilicon emitter is doped with arsenic whereas the bottom part is undoped. The diffusion coefficient of arsenic down grain boundaries is a factor of 10^4 times higher [8] than that in bulk silicon, and hence the arsenic diffuses very rapidly down the grain boundaries and reaches the polysilicon/silicon interface after the first few seconds of the anneal. At this point in time, in the bottom part of the layer almost all of the arsenic is segregated at the grain boundaries and the centres of the grains are undoped. In the remainder of the anneal the arsenic slowly diffuses from the grain boundaries into the grain interiors and from the polysilicon/silicon interface into the single-crystal emitter. The diffusion into the single-crystal emitter is much slower than the grain boundary diffusion, and hence it is possible to accurately control the depth of the emitter/base junction by choosing an appropriate anneal temperature and time.

Figure 6.5 illustrates the typical sequence of doping profiles obtained at different times during an emitter anneal at 1025°C. Diffusion of arsenic down grain boundaries is extremely fast, and hence after the first few seconds of the anneal, the arsenic will reach the polysilicon/silicon interface. A small peak in the arsenic concentration will be present at the polysilicon/silicon interface due to arsenic segregation. This segregation peak occurs because the interface behaves like a large grain boundary.

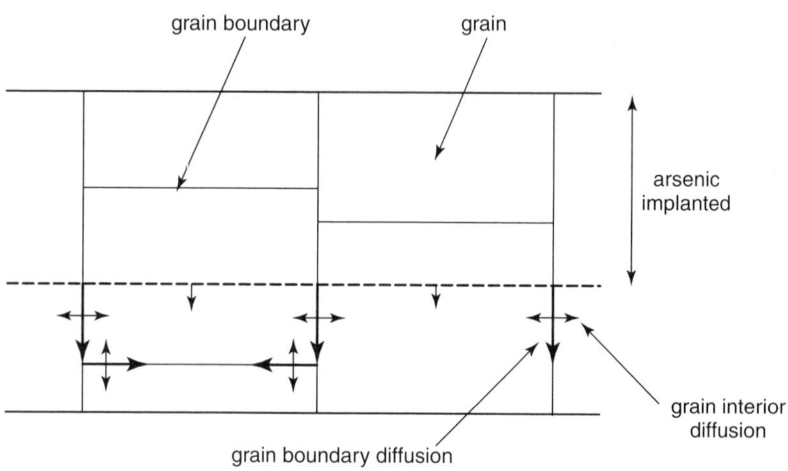

Figure 6.4 Schematic illustration of arsenic diffusion in polysilicon emitters

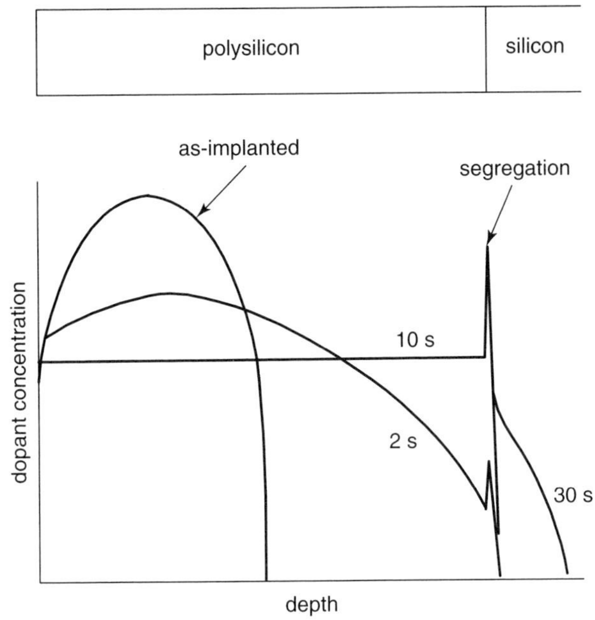

Figure 6.5 Schematic illustration of the sequence of doping profiles obtained at different times during the anneal of a polysilicon emitter. A logarithmic concentration scale is assumed so that full details of the profiles can be seen

After around 10 seconds, the arsenic concentration in the polysilicon will be approximately uniform because of the high grain boundary diffusion coefficient of arsenic. It should be noted that arsenic segregated at grain boundaries is not electrically active, so only a fraction of the arsenic in the polysilicon will be electrically active, typically between 25 and 50% [9]. After about thirty seconds, dopant penetration into the single-crystal silicon will be seen. The majority of the arsenic in the single-crystal silicon will be electrically active, so the arsenic concentration at the single-crystal side of the interface will be lower than that at the polysilicon side, as illustrated in Figure 6.5. This difference in concentrations gives a guide to the fraction of electrically active arsenic in the polysilicon.

Grain growth generally occurs during high temperature annealing of polysilicon. This leads to a decrease in the density of grain boundaries with anneal time, release of segregated dopant from the grain boundaries and a consequent decrease in polysilicon sheet resistance with time. Grain growth will also influence the dopant diffusion in the polysilicon, since there will be fewer grain boundaries to provide rapid dopant diffusion paths to the interface. Models have been developed [10] that incorporate all of the above mechanisms and allow the simulation of dopant diffusion in polysilicon in software packages such as SUPREM.

6.4 INFLUENCE OF THE POLYSILICON/SILICON INTERFACE

A further complication in the behaviour of polysilicon emitters is the polysilicon/silicon interface. Silicon oxidizes very readily in oxygen, and a thin native oxide layer forms on the surface of silicon wafers even at room temperature. This means that a thin oxide layer is invariably present at the polysilicon/silicon interface in polysilicon emitters. The thickness of this interfacial oxide layer depends on the ex situ surface cleans used prior to wafer insertion in the polysilicon deposition furnace, on the type of deposition system used and on any in situ cleans done inside the deposition system prior to deposition. For the case of an ex situ hydrofluoric acid (HF) etch and polysilicon deposition in a Low Pressure Chemical Vapour Deposition (LPCVD) furnace, the interfacial oxide thickness is typically 0.4 nm [11], and for an RCA clean is typically 1.4 nm [11]. While these might be considered to be negligibly thin oxide layers, they nevertheless have a strong influence on the base current of transistors with polysilicon emitters.

The electrical effect of an interfacial oxide is illustrated in Figure 6.6, which compares Gummel plots for transistors with polysilicon emitters

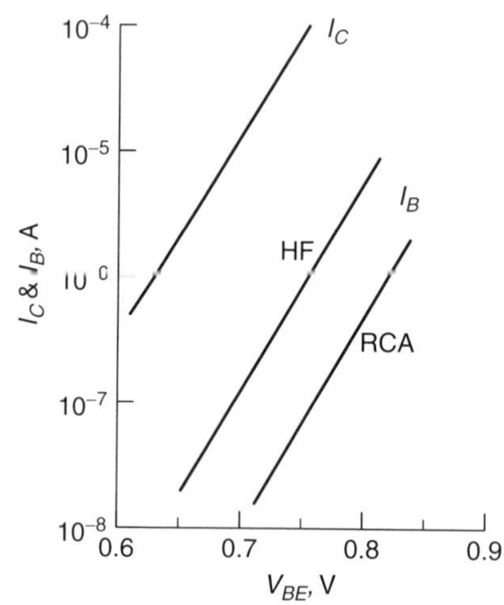

Figure 6.6 Gummel plots illustrating the influence of the surface clean, and hence interfacial layer thickness, on the base current of transistors with polysilicon emitters

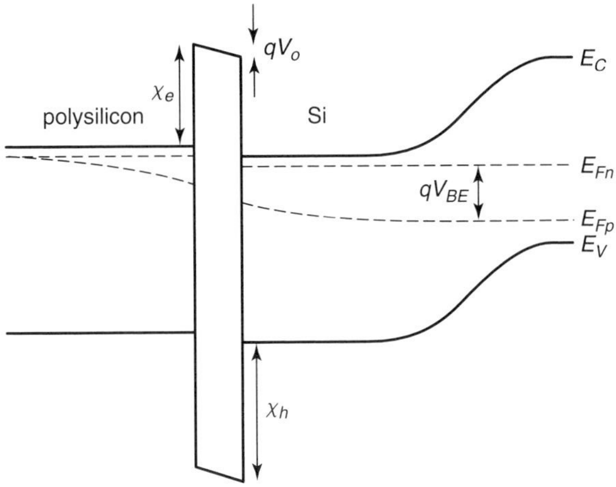

Figure 6.7 Band diagram for a polysilicon emitter

created using an HF etch and an RCA clean. The thicker interfacial layer created by the RCA clean leads to a suppression of the base current, and consequently a much higher gain. Experimental measurements have shown that polysilicon emitters produced using an HF etch have gains enhanced by a factor of 2 or 3 compared with similar transistors with ion implanted emitters [4,11]. Polysilicon emitters produced using an RCA clean have gain enhancements of a factor of 10 or more [4,11].

The effect of the interfacial oxide layer on the base current of polysilicon emitter bipolar transistors can be explained by the band diagram in Figure 6.7. The presence of the interfacial oxide layer creates a potential barrier χ_h for holes injected from the base into the emitter. The primary mechanism by which holes traverse this barrier to reach the emitter contact is tunnelling [2,4,12]. The hole current, and hence the base current, is therefore determined by the tunnelling properties of the interfacial oxide, in particular by the interfacial layer thickness and the effective barrier height for holes χ_h. Since the tunnelling current decreases exponentially with the interfacial oxide thickness, it is clear that the base current will depend strongly on the physical properties of the polysilicon/silicon interface.

6.5 BASE CURRENT IN POLYSILICON EMITTERS

As discussed above, the presence of an interfacial oxide layer gives rise to a lower value of base current than would be expected and hence

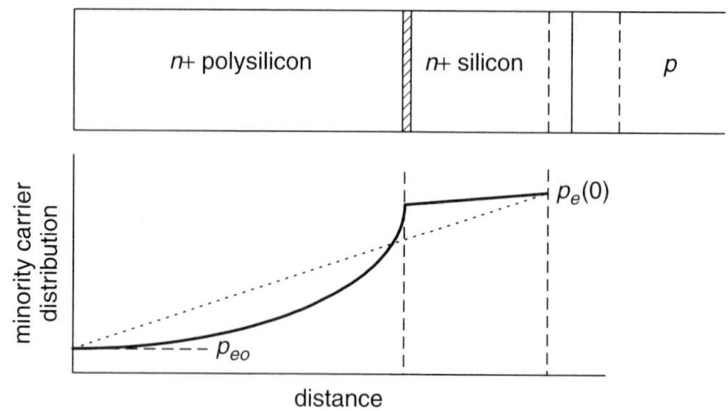

Figure 6.8 Hole distribution in a polysilicon emitter with an interfacial layer, and for comparison the hole distribution in a single-crystal emitter

a higher value of gain. The effect of the interfacial layer on the hole distribution in the emitter is illustrated in Figure 6.8. The gradient of the hole distribution in the single-crystal emitter is smaller than that for the equivalent single-crystal emitter because of the suppression of hole transport across the polysilicon/silicon interface by the interfacial layer. In the polysilicon part of the emitter, the hole distribution will depend on the structure of the polysilicon layer. The grain boundaries in the polysilicon contain trapping states and hence would be expected to decrease the diffusion length compared with single-crystal silicon. In this case, the hole concentration would drop sharply on entering the polysilicon, as illustrated in Figure 6.8.

The base current can be easily derived by treating the polysilicon layer as a metal contact, but with a lower value of surface recombination velocity. For a metal contact, the surface recombination is very high ($1 \times 10^6\,\mathrm{cm\,s^{-1}}$), so all holes recombine at the metal contact. The hole concentration adjacent to the metal contact is then equal to the equilibrium hole concentration p_{eo}. For a polysilicon emitter, Figure 6.8 shows that the hole concentration at the polysilicon/silicon interface is considerably higher than the equilibrium value p_{eo}. This is equivalent to saying that the effective recombination velocity at the polysilicon/silicon interface is much lower than would be obtained if the polysilicon layer was replaced by a metal contact. It is therefore possible to model the polysilicon emitter using an effective surface recombination velocity at the polysilicon/silicon interface S_{EFF}, as illustrated in Figure 6.9. In this case, the effective surface recombination velocity S_{EFF} is defined

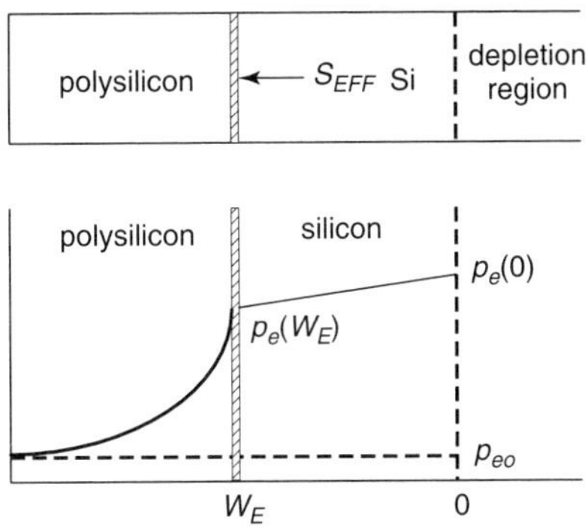

Figure 6.9 Effective surface recombination velocity S_{EFF} for a polysilicon emitter

as follows:

$$J_p(W_E) = qS_{EFF}(p_e(W_E) - p_{eo}) \tag{6.1}$$

where $J_P(W_E)$ and $p_e(W_E)$ are the hole current density and hole concentration at the polysilicon/silicon interface.

The hole concentration at the polysilicon/silicon interface can be written in terms of S_{EFF} as:

$$p_e(W_E) = \frac{J_p(W_E) + qS_{EFF}p_{eo}}{qS_{EFF}} \tag{6.2}$$

If we assume that the single-crystal emitter is shallow, there will be little recombination in the single-crystal emitter and the hole distribution will be linear, as shown in Figure 6.9 and $J_p(W_E) = J_p(0) = J_p$. The gradient of the hole distribution is then:

$$\begin{aligned} \text{gradient} &= -\frac{p_e(0) - p_e(W_E)}{W_E} \\ &= -\frac{p_{eo}\exp\dfrac{qV_{BE}}{kT} - p_e(W_E)}{W_E} \end{aligned} \tag{6.3}$$

The hole current density is a diffusion current, which is given by:

$$\begin{aligned} J_p &= -qD_{pe}\frac{\mathrm{d}p_e}{kT} \\ &= -qD_{pe} \times \text{gradient} \end{aligned} \tag{6.4}$$

Substituting for the gradient from equation (6.3) and $p_e(W_E)$ from equation (6.2) gives:

$$J_p = \frac{qp_{eo}}{\dfrac{W_E}{D_{pe}} + \dfrac{1}{S_{EFF}}}\left(\exp\frac{qV_{BE}}{kT} - 1\right) \qquad (6.5)$$

Substituting for p_{eo} using equation (2.25) and including heavy doping effects leads to the following equation for the base current of a polysilicon emitter:

$$I_B = \frac{qAn_{io}^2}{\dfrac{N_{deff}W_E}{D_{pe}} + \dfrac{N_{deff}}{S_{EFF}}}\exp\frac{qV_{BE}}{kT} \qquad (6.6)$$

where it has been assumed that $V_{BE} \gg kT/q$. When S_{EFF} is large (thin interfacial oxide layer), the second term in the denominator can be neglected and equation (6.6) reduces to the standard equation for the base current of a bipolar transistor, as given in equation (3.18). When S_{EFF} is small (thick interfacial oxide layer), the first term in the denominator can be neglected and in this case, the base current is proportional to S_{EFF}.

6.6 EFFECTIVE SURFACE RECOMBINATION VELOCITY

Theoretical models have been derived for the effective recombination velocity at the polysilicon/silicon interface, S_{EFF}, which incorporate all the physical mechanisms occurring in the polysilicon emitter [13]. In general, S_{EFF} can be expressed as a combination of effective recombination velocities [13]:

$$S_{EFF} = S_I + \frac{1}{\dfrac{1}{T_I} + \dfrac{1}{S_I + S_P}} \qquad (6.7)$$

where T_I models hole tunnelling through the interfacial oxide layer, S_I models recombination at the interface between the interfacial layer and the single-crystal emitter and between the interfacial layer and the polysilicon, and S_P models hole transport through the polysilicon layer. The tunnelling term T_I has a strong effect on S_{EFF} and is given by [13]:

$$T_I = \sqrt{\frac{kT}{2\pi m_h^*}} \frac{\exp(-b_h)}{(1 - c_h kT)} \tag{6.8}$$

$$c_h = \frac{2\pi\delta}{h}\sqrt{\frac{2m_h^*}{\chi_h}} \tag{6.9}$$

$$b_h = \frac{4\pi\delta}{h}\sqrt{2m_h^*\chi_h} \tag{6.10}$$

where δ is the thickness of the interfacial oxide layer, and m_h^* is the effective mass for holes. A full derivation of equations (6.7)–(6.10) is given in [13] and [14].

S_P depends on the grain size in the polysilicon, and typically varies between about 2100 ms^{-1} for a single grain in the polysilicon layer and 1500 ms^{-1} for three grains or more [13]. S_I depends on the structure and quality of the interfacial oxide layer and typically has a value of 15 ms^{-1} [13]. The parameter that has the biggest effect on the value of S_{EFF} is χ_h; modelling of the Gummel plots of *npn* polysilicon emitter bipolar transistors has shown that good agreement between measured and modelled values of base current can be obtained when $\chi_h \approx 1.1$ eV [15].

The variation of S_{EFF} with interfacial oxide thickness is schematically plotted in Figure 6.10, where it has been assumed that $S_I = 15\,ms^{-1}$, $S_P = 1500\,ms^{-1}$, $\chi_h = 1.1$ eV [15], $T = 300$ K and $m_h^* = m_o$, the free electron mass. For thick interfacial oxide layers, hole transport through the interfacial oxide is strongly suppressed, so T_I is small and in this case equation (6.7) shows that $S_{EFF} \approx S_I$. In this part of the characteristic, S_{EFF} is therefore limited by recombination at the interface between the single-crystal emitter and the interfacial oxide. This recombination occurs at dangling bonds at the interfacial oxide/silicon interface, and limits the achievable gain in polysilicon emitters with a thick interfacial oxide. Research has shown that fluorine implanted into the polysilicon emitter is very effective in passivating the dangling bonds at the polysilicon/silicon interface, and hence in reducing recombination [16]. A fluorine implant into the polysilicon emitter therefore provides a way of improving the gain when the interfacial oxide is thick.

For intermediate values of interfacial oxide thickness (between about 0.3 and 0.7 nm), hole transport through the interfacial oxide is not so strongly suppressed, so S_{EFF} becomes very sensitive to the interfacial layer thickness. In this region of the characteristic in Figure 6.10 S_{EFF} is

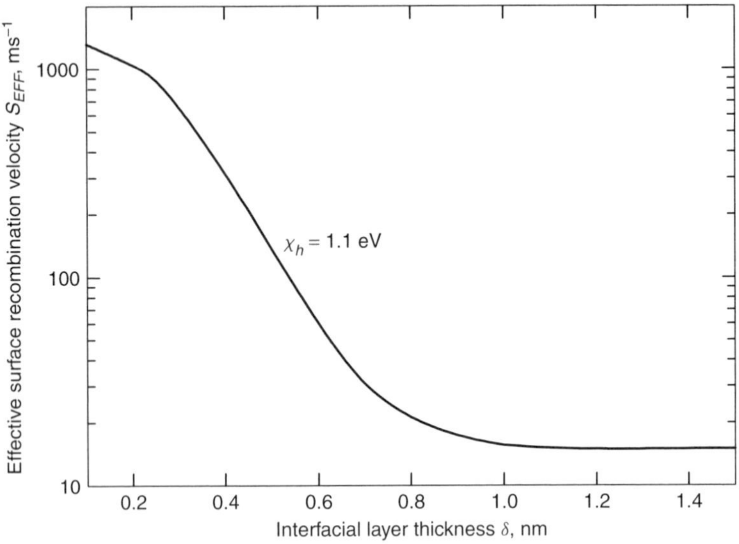

Figure 6.10 Variation of effective surface recombination velocity S_{EFF} with interfacial oxide thickness (reproduced with permission from the American Institute of Physics [11])

dominated by tunnelling through the interfacial oxide layer and hence the base current is very sensitive to the interfacial layer thickness. This situation is very undesirable in a production transistor because it is very difficult to control the gain of the transistor both across a wafer and from wafer to wafer.

For thin interfacial oxide layers (less than 0.3 nm), tunnelling through the interfacial layer becomes increasingly less important as the interfacial oxide thickness is decreased and the value of S_{EFF} approaches S_P. In this case, transport in the polysilicon layer dominates, and parameters like the thickness of the polysilicon layer, the grain size and the grain structure influence the value of S_{EFF}. Grain boundaries are present in the polysilicon layer and also at the interface between the polysilicon and interfacial oxide layer, which can be considered as a large pseudo-grain boundary. Grain boundaries contain a high density of defects and dangling bonds, and hence can act as recombination centres for the minority carrier holes. It has been shown experimentally that arsenic segregation at grain boundaries passivates the dangling bonds at the grain boundary, and hence decreases the base current [17]. A decrease in base current by approximately 40% was obtained in polysilicon emitters when arsenic was deliberately segregated to the grain boundaries. It has also been experimentally demonstrated that segregation at the pseudo-grain boundary at the polysilicon/silicon interface also decreases the base current [18].

In practice, it is generally not necessary to calculate the effective surface recombination velocity S_{EFF} using equations (6.7)–(6.10) because S_{EFF} is generally used as a fitting parameter to model the measured base current. Nevertheless, these equations are useful in providing insight into the factors that influence the gain of a polysilicon emitter bipolar transistor.

6.7 EMITTER RESISTANCE

The band diagram in Figure 6.7 indicates that the interfacial layer gives rise to a potential barrier in the conduction band as well as the valence band. We have already seen that the potential barrier in the valence band has the effect of suppressing hole transport through the interfacial oxide because the holes have to tunnel through the barrier to reach the emitter contact. Holes are minority carriers in the emitter and hence the electrical effect of the potential barrier in the valence band is to decrease the base current, as shown in Figure 6.6. Electrons are majority carriers in the emitter and hence we would expect the electrical effect of the potential barrier in the conduction band on electron flow to be different than that seen for holes. This is indeed the case, and as might be imagined, the potential barrier in the conduction band has the effect of increasing the series resistance of the emitter. Polysilicon emitters containing an interfacial oxide layer therefore have increased emitter resistance.

The emitter resistance is determined by tunnelling through the interfacial oxide layer, and can be approximated by [14,15]:

$$R_E = \frac{(1 - c_e kT)}{qAc_e T^2 A_e} \exp b_e \exp \frac{E_C - E_F}{kT} \tag{6.11}$$

where c_e and b_e are given by equations analogous to equations (6.9) and (6.10) and A_e is the Richardson constant given by:

$$A_e = \frac{4\pi q m_e^* k^2}{h^3} \tag{6.12}$$

Equation (6.12) shows that the emitter resistance increases strongly with interfacial layer thickness, δ and electron effective barrier height χ_e through the term b_e. Modelling of the Gummel plots of *npn* polysilicon emitter bipolar transistors has shown that good agreement between the measured emitter resistance and the model can be obtained when $\chi_e \approx 0.4\,\text{eV}$ [15].

Emitter resistance is undesirable in practical bipolar transistors because it degrades both the current carrying capability of the bipolar transistor and the transconductance, as given by:

$$g_m = \frac{\mathrm{d}I_C}{\mathrm{d}V_{BE}} \tag{6.13}$$

Since the collector current varies exponentially with base/emitter voltage, as shown in equation (3.19), the bipolar transistor should have a very high transconductance. However, if emitter resistance is present, a linear dependence of collector current on base/emitter voltage is obtained that drastically degrades the transconductance. In polysilicon emitters, there is therefore a trade-off between current gain and emitter resistance. Transistors with thick interfacial oxide layers have the advantage of very high gains, but the disadvantage of very high values of emitter resistance. Transistors with thin interfacial oxide layers have the advantage of low values of emitter resistance, but the disadvantage of lower values of current gain. As is the case for many engineering situations, it is therefore necessary to strike a balance between the value of emitter resistance and current gain.

6.8 DESIGN OF PRACTICAL POLYSILICON EMITTERS

6.8.1 Break-up of the Interfacial Oxide Layer and Epitaxial Regrowth

The band diagram in Figure 6.7 is somewhat idealized in that it assumes that the interfacial oxide layer is uniform in thickness across the emitter of the transistor. While this is the case immediately after deposition of the polysilicon layer, high temperature anneals have a strong influence on the interfacial oxide layer, and in practice can lead to the break-up of the interfacial oxide.

Figure 6.11 shows a schematic illustration of the break-up of the interfacial oxide layer interface during high-temperature anneal. After deposition, the interfacial oxide layer is uniform in thickness and continuous [11], as illustrated in Figure 6.11(a). The thickness of the interfacial oxide depends on the ex situ clean carried out prior to insertion of the wafers in the polysilicon deposition system, and also on the type of deposition system [6]. For an ex situ HF etch and deposition in a furnace

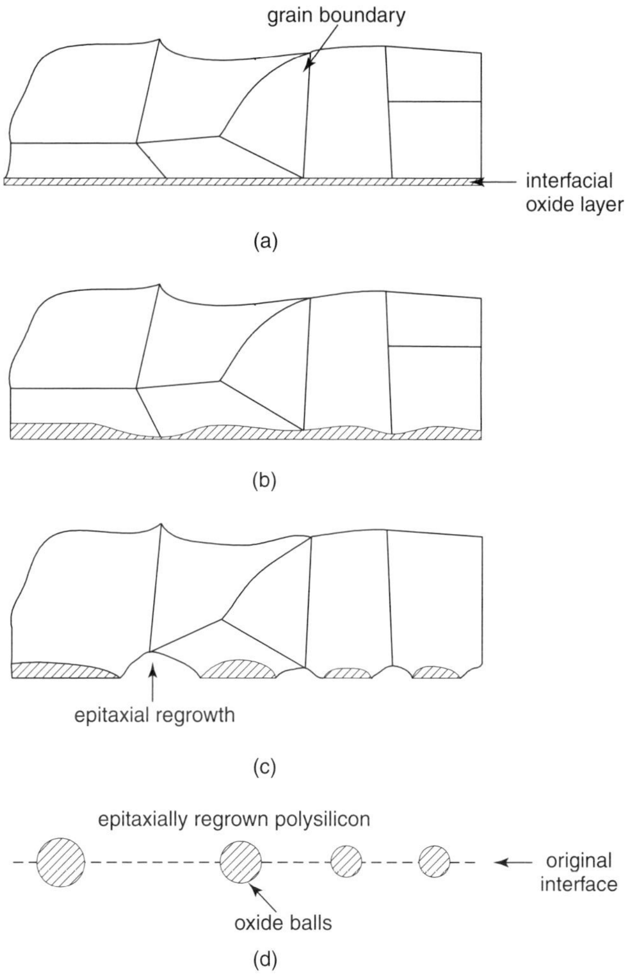

Figure 6.11 Schematic illustration of interfacial oxide break-up during high temperature anneal of a polysilicon emitter (after [11])

LPCVD system, the interfacial oxide layer is typically around 0.4 nm thick, whereas for an RCA clean and deposition in a furnace LPCVD deposition system the interfacial oxide is typically 1.4 nm thick [11]. For deposition in cluster tools, thinner interfacial oxide layers are obtained [6] because of the better vacuum conditions during wafer insertion into the system and during growth.

A high-temperature anneal ($\geq$950°C) causes a degradation in the uniformity of the interfacial layer, with some parts of the layer becoming thicker and others thinner, as illustrated in Figure 6.11(b). Annealing for longer times (or at higher temperatures) causes further stressing of the interfacial oxide until holes appear in the interfacial oxide as shown in Figure 6.11(c). At this point, local epitaxial regrowth of the polysilicon occurs through the holes in the interfacial oxide layer. The epitaxial regrowth tends to occur at points where grain boundaries intersect the interfacial oxide layer. Research has shown [17] that a high concentration of arsenic in the polysilicon emitter aids break-up of the interfacial oxide. This suggests that the diffusion of arsenic down grain boundaries to the interfacial oxide is one of the mechanisms that drives the break-up of the interfacial oxide. Research has also shown that fluorine implanted into the polysilicon emitter also aids break-up of the interfacial oxide layer [19]. Grain growth also occurs during high-temperature anneal, and there is a tendency for grains to become more columnar in shape, as illustrated in Figure 6.11(c), particularly when the doping concentration in the polysilicon is high.

Anneals at very high temperatures ($\geq$1000°C) lead to balling of the interfacial oxide and the complete epitaxial regrowth of the polysilicon, as illustrated in Figure 6.11(d). One of the driving forces for the break-up of the interfacial layer is believed to be minimization of the energy associated with the polysilicon/silicon interface [11]. In general, the energy of the interface is minimized when the surface area of the interfacial oxide is minimized. This condition is, of course, met when the oxide forms itself into spherical balls at the interface.

Table 6.2 illustrates how the interfacial oxide break-up influences the base current of transistors with polysilicon emitters. The devices were subjected to a short, high-temperature interface anneal at a temperature in the range 800–1100°C after polysilicon deposition but prior to the emitter implant [11]. This interface anneal causes the interfacial layer to break up, with the extent of the break-up being determined by the

Table 6.2 Effect of anneal temperature on base current and interfacial oxide break-up in polysilicon emitters (data taken from [11] reproduced with permission from the American Institute of Physics)

Interface anneal Temperature,°C	Base saturation current, A x 10^{-20}		Interfacial oxide break-up, %	
	RCA clean	HF etch	RCA clean	HF etch
none	3	30	0	20
800	3	30	0	20
900	4	40	0	20
950	18	85	30	–
1000	84	170	70	oxide balls
1100	300	–	oxide balls	–

anneal temperature. The results in Table 6.2 show very little change in base current until a temperature of 950°C, at which point the base current increases dramatically. For devices given an ex situ RCA clean prior to polysilicon deposition, the base current increases by a factor of 6 as the interfacial oxide break-up increases from zero to 30%. For devices given an HF etch the increase in base current is by a factor of 2.8. In transistors given an RCA clean, balling of the interfacial oxide occurs after anneal at 1100°C, at which point the base current has increased by a factor of 100 compared with the control transistor given no interface anneal. In transistors given an HF etch, balling occurs after an interface anneal at 1000°C, and this corresponds to a base current increase by a factor of 5.7 compared with the control transistor.

The above results clearly indicate that the polysilicon/silicon interface can be engineered to give different combinations of emitter resistance and current gain. For example, if high gains are of paramount importance then the process should include an RCA treatment, the polysilicon should not be too heavily doped and the emitter drive-in should be carried out at a temperature of 900°C or lower. Under these circumstances a continuous and uniform interfacial layer will result, and tunnelling of holes through the interfacial oxide will lead to suppression of the base current and hence the required high gains. In general, low values of emitter resistance are of paramount importance because of the degradation in circuit performance that is obtained if the emitter resistance is too high, so some method of reducing the emitter resistance is needed.

6.8.2 Epitaxially Regrown Emitters

Extremely low values of emitter resistance can be achieved if the polysilicon is epitaxially regrown during the emitter anneal. Epitaxially regrown polysilicon emitters are electrically identical to conventional single-crystal emitters. The base current and gain are given by equations analogous to equations (3.15) and (3.17):

$$I_B = \frac{qAD_{pe}n_{io}^2}{(W_E + W_{POL})N_{deff}} \exp\frac{qV_{BE}}{kT} \tag{6.14}$$

$$\beta = \frac{D_{nb}(W_E + W_{POL})N_{deff}}{D_{pe}W_BN_{aeff}} \tag{6.15}$$

where the emitter width is given by the sum of the penetration into the single-crystal silicon emitter W_E and the polysilicon layer thickness W_{POL}.

Epitaxial regrowth of the polysilicon has advantages for production, since the base current is not controlled by the interfacial oxide thickness, but by the polysilicon thickness and the arsenic penetration depth into the single-crystal emitter. Good control and reproducibility of the base current and current gain can therefore be achieved. To ensure epitaxial regrowth of the polysilicon, an HF etch should be used prior to polysilicon deposition and the arsenic concentration in the polysilicon layer should be high to encourage interfacial oxide break-up. Table 6.2 shows that interfacial oxide break-up occurs more readily when the emitter anneal is performed at a high temperature, so a short rapid thermal anneal at a temperature above 1000°C should be used in preference to a longer furnace anneal at a lower temperature. One disadvantage of epitaxially regrown polysilicon emitters is that the high temperature needed to break up the interfacial oxide layer and epitaxially regrow the polysilicon is not compatible with ultra-shallow emitter/base junctions.

In *n*-type polysilicon layers, epitaxial regrowth is accompanied by a decrease in the sheet resistance of the polysilicon-on-silicon layer [20]. This behaviour is related to the segregation of arsenic and phosphorus to grain boundaries. The segregated arsenic is electrically inactive and hence does not contribute to the sheet resistance of the polysilicon layer. When epitaxial regrowth occurs, the arsenic is released from the grain boundaries and hence is able to diffuse to find substitutional lattice sites, where it becomes electrically active. In practice epitaxial regrowth leads to a decrease in polysilicon-on-silicon sheet resistance by a factor of approximately two [20]. This phenomenon is very useful for process control, since the presence of epitaxial regrowth can be detected from a simple measurement of the sheet resistance of polysilicon-on-silicon process monitor structures.

6.8.3 Trade-off between Emitter Resistance and Current Gain in Polysilicon Emitters

If ultra-shallow emitter/base junctions are required, low emitter anneal temperatures have to be used, which make it difficult to epitaxially regrow the polysilicon layer. In this situation, the trade-off between the emitter resistance and the current gain is of critical importance. Research has shown that the emitter resistance decreases much more quickly with interfacial oxide break-up than the base current increases [21]. This situation is illustrated schematically in Figure 6.12. The majority of the drop in emitter resistance occurs in the early stages of interfacial

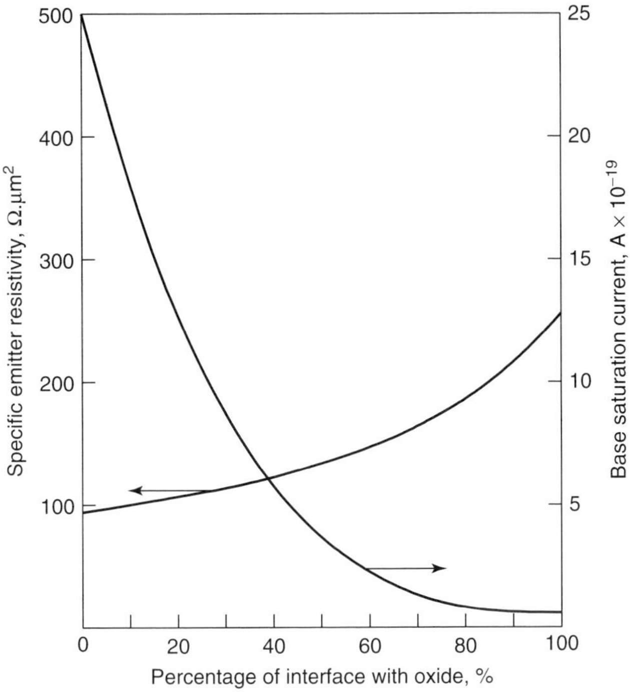

Figure 6.12 Trade-off between emitter resistance and base current in a polysilicon emitter (reprinted with permission from [21])

oxide break-up when the first holes form in the interfacial oxide. In Figure 6.12, this is represented by the large fall in emitter resistance when the fraction of the interface covered with oxide decreases from 100% to 80%. The physical explanation for this result is illustrated in Figure 6.13. When gaps in the interfacial occur, the majority carrier electron current is diverted laterally to find the path of least resistance through the gaps in the interfacial oxide. The most favourable situation is when there are a large number of very small gaps, so that the lateral diversion of current is small and hence the lateral series resistance small. This is precisely the situation that occurs in a polysilicon emitter during the initial stages of interfacial oxide break-up, since the break-up tends to occur at the points where grain boundaries intersect the interfacial oxide layer.

The increase in base current with increasing interfacial oxide break-up is relatively slow, as illustrated in Figure 6.12. The physical explanation for this result is again demonstrated in Figure 6.13. The minority carrier hole flow in the single-crystal part of the emitter is by diffusion, and

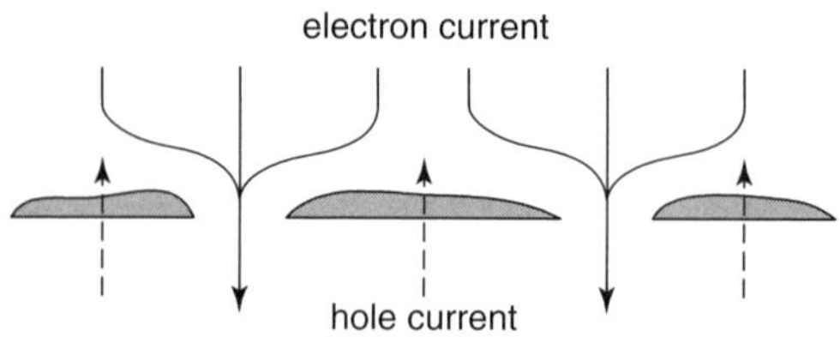

Figure 6.13 Schematic illustration showing the electron and hole current flow through gaps in the interfacial oxide layer

transport across the interfacial oxide layer by tunnelling, and hence there is little lateral flow of holes. In the early stages of interfacial oxide break-up, the minority carrier hole current is therefore primarily determined by the thickness of the interfacial layer, rather than the gaps in the interfacial oxide. Since thinning of the interfacial oxide occurs slowly in the initial stages of interfacial oxide break-up, the increase in base current is also slow. In the later stages of interfacial oxide break-up, when less than 50% of the interface is covered with oxide, the gaps in the oxide increasingly dominate the hole current and hence the base current increases sharply, as shown in Figure 6.12.

To produce polysilicon emitters in which the polysilicon remains polycrystalline it is necessary to tightly control the structure of the interfacial oxide. The polysilicon/silicon interface needs to be broken-up sufficiently to give low values of emitter resistance but not so much that the benefits of higher gain are lost. Polysilicon deposition in a cluster tool offers better control of the interfacial oxide than deposition in an LPCVD furnace, because of the cleaner growth environment [6,22,23]. An ex situ HF interface treatment is therefore generally used in combination with deposition in a cluster tool.

A number of process variables influence the polysilicon structure, including deposition temperature, emitter implant dose, and emitter anneal conditions. When the deposition temperature is 580°C or below, the polysilicon is deposited in the form of amorphous silicon, whereas temperatures above 580°C give polycrystalline silicon. The amorphous silicon is converted to polycrystalline form during subsequent high-temperature treatments, but the end result is a larger grain size [24]. The advantages of a larger grain size are lower resistances and films with a high degree of surface smoothness. In addition, a larger grain size gives fewer grain boundaries, less dopant diffusion to the interface and hence less interfacial oxide break-up. Emitter implant dose determines the dopant concentration in the polysilicon and high dopant concentrations enhance grain growth [25] and interfacial oxide break-up. Similarly high emitter anneal temperatures also increase grain growth and interfacial oxide break-up. If the polysilicon layer is to remain polycrystalline, very high emitter implant doses and high emitter anneal temperatures should be avoided.

6.8.4 Emitter Plug Effect and in situ Doped Polysilicon Emitters

In practice, polysilicon emitters are formed in emitter windows in oxide layers covering the surface of the silicon wafer. Polysilicon deposition is very conformal, which means that the polysilicon layer will follow the contour of the layers onto which it is deposited, as shown in Figure 6.14(a). In this situation, the polysilicon layer at the perimeter (P) of the emitter is thicker than that in the centre (C). If the emitter is doped by ion implantation, the dopant will have further to diffuse to reach the interface at the perimeter of the emitter than in the centre. Consequently, the emitter/base junction depth will be shallower at the perimeter of the emitter than in the centre. This phenomenon is known

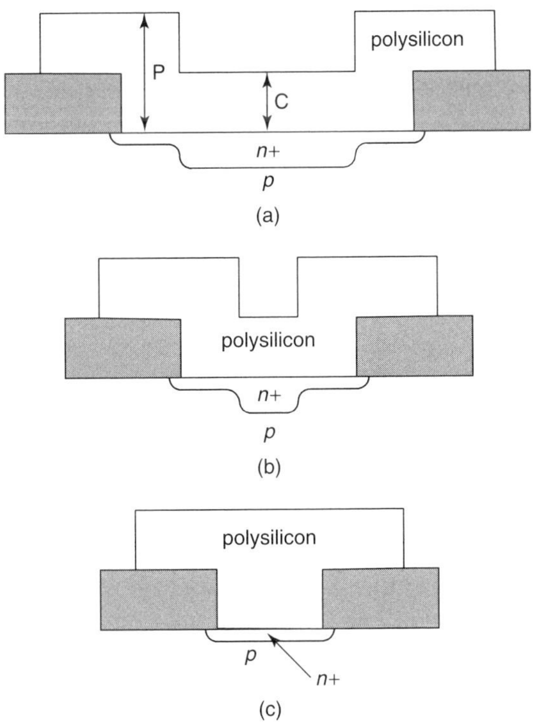

Figure 6.14 Schematic illustration of the emitter plug effect for a large (a), intermediate (b) and small (c) emitter window

as the emitter plug effect, and it causes problems in controlling the emitter/base junction depth. If the emitter/base junction depth is very shallow, as will be the case in very high-speed devices, the emitter/base depletion region may penetrate to the polysilicon/silicon interface at the perimeter of the emitter. The large number of trapping states at the interface between the interfacial oxide and the single-crystal emitter will give rise to recombination in the emitter/base depletion region and hence non-ideal base characteristics, as shown in Figure 4.1.

The emitter plug effect also makes it difficult to scale the emitter/base junction depth, as illustrated in Figures 6.14(b) and (c). As the emitter window size is reduced, the thicker polysilicon layer at the emitter perimeter takes up an increasing fraction of the emitter window until eventually it is completely plugged, as illustrated in Figure 6.14(c). The severity of the emitter plug effect depends on the thickness of the oxide layer at the perimeter of the emitter window. As will be seen in Chapter 9, in double polysilicon bipolar processes the thickness of this oxide can approach 1 μm, and in this case the emitter plug effect is very severe.

The problems caused by the emitter plug effect can be solved by doping the polysilicon during deposition instead of using ion implantation [26]. If the polysilicon is in situ doped, the dopant concentration adjacent to the polysilicon/silicon interface is uniform across the whole of the emitter window. The in situ doped polysilicon layer therefore acts as a very ideal diffusion source, and hence the emitter/base junction depth is uniform across the emitter window. The use of in situ doped polysilicon is highly advisable when the emitter window size is smaller than twice the polysilicon thickness.

6.9 *pnp* POLYSILICON EMITTERS

As will be discussed in Chapter 9, *pnp* polysilicon emitters are used in complementary bipolar processes, and in this case arsenic is used as the base dopant and boron as the emitter dopant. Boron diffusion in polysilicon is a little different to arsenic and phosphorus diffusion because boron diffusion down grain boundaries is slower. For arsenic, grain boundary diffusion is a factor of approximately 10^4 times higher than bulk diffusion, whereas for boron this factor is only 100 [27]. For short anneals, the flat profiles that are characteristic of arsenic diffusion (Figure 6.6) are not seen. The slower grain boundary diffusion of boron leads to a decrease in boron concentration as the interface

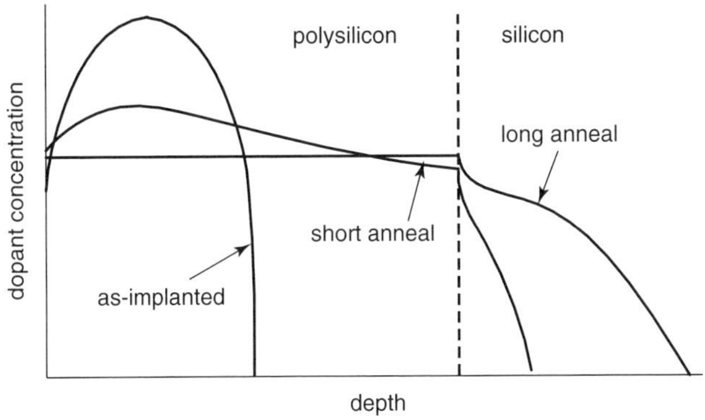

Figure 6.15 Schematic illustration of boron diffusion in polysilicon for *pnp* polysilicon emitters

is approached, as illustrated in Figure 6.15. The boron concentration at the polysilicon/silicon interface is therefore lower than would be obtained for arsenic for the same emitter/base junction depth. This is a disadvantage for shallow emitter/base junction formation because there is more penetration of the emitter/base depletion width into the single-crystal emitter if the doping is lower. The boron concentration at the polysilicon/silicon interface can be increased by annealing for longer or at a higher temperature, but in this case a much deeper emitter/base junction is obtained, as illustrated in Figure 6.15. Furthermore, boron does not segregate to grain boundaries like arsenic and phosphorus, so no peak is observed in the boron concentration at the polysilicon/silicon interface. The epitaxial regrowth of boron doped polysilicon is therefore not accompanied by a decrease in sheet resistance, as is the case for arsenic and phosphorus.

The band diagram in Figure 6.7 suggests that the electrical behaviour of *pnp* polysilicon emitters would be different to that of *npn*. In *npn* transistors, χ_h is bigger than χ_e and hence a big decrease in base current is obtained with a small increase in emitter resistance. If this band diagram was directly applicable to *pnp* transistors, we would expect to see a small decrease in base current with a large increase in emitter resistance. Fortunately, in practice this behaviour is not seen and the electrical performance of *pnp* polysilicon emitters is very similar to that of *npn* transistors, with a large suppression of current gain and a small increase in emitter resistance [28,29]. A model has been proposed for this behaviour which assumes that the interfacial oxide layer behaves like a wide bandgap semiconductor rather than an insulator [28,29].

REFERENCES

[1] J. Graul, A. Glasl and H. Murrmann, 'High-performance transistors with arsenic-implanted polysil emitters', *IEEE Jnl Solid State Circuits*, **11**, 491 (1976).

[2] H.C. De Graaff and J.G. De Groot, 'The SIS tunnel emitter: a theory for emitters with thin interfacial layers', *IEEE Trans.* **26**, 1771 (1979).

[3] T.H. Ning and R.D. Isaac, 'Effect of emitter contact on current gain of silicon bipolar devices', *IEEE Trans. Electron. Devices*, **27**, 2051 (1980).

[4] P. Ashburn and B. Soerowirdjo, 'Comparison of experimental and theoretical results on polysilicon emitter bipolar transistors', *IEEE Trans. Electron. Devices*, **31**, 853 (1984).

[5] D. Tang and P. Solomon, 'Bipolar transistor design for optimized power-delay logic circuits', *IEEE Jnl Solid State Circuits*, **14**, 679 (1979).

[6] A.I.A. Rahim, C.D. Marsh, P. Ashburn and G.R. Booker, 'Impact of ex-situ and in-situ cleans on the performance of bipolar transistors with low thermal budget in-situ phosphorus-doped polysilicon emitter contacts', *IEEE Trans. Electron Devices*, **48**, 2506 (2001).

[7] M.M. Mandurah, K.C. Saraswat and C.R. Helms, 'Dopant segregation in polycrystalline silicon', *Jnl Appl. Phys*, **51**, 5755 (1980).

[8] B. Swaminathan, K.C. Saraswat, R.W. Dutton and T.I. Kamins, 'Diffusion of arsenic in polycrystalline silicon', *Appl. Phys. Lett.* **40**, 795 (1982).

[9] A. Cuthbertson and P. Ashburn, 'An investigation of the trade-off between enhanced gain and base doping in polysilicon emitter bipolar transistors', *IEEE Trans. Electron. Devices*, **32**, 2399 (1985).

[10] A.G. O'Neill, C. Hill, J. King and C. Please, 'A new model for the diffusion of arsenic in polycrystalline silicon', *Jnl Appl. Phys.* **64**, 167 (1988).

[11] G.R. Wolstenholme, N. Jorgensen, P. Ashburn and G.R. Booker, 'An investigation of the thermal stability of the interfacial oxide in polycrystalline silicon emitter bipolar transistors by comparing device results with high-resolution electron microscopy observations', *Jnl Appl. Phys.*, **61**, 225 (1987).

[12] I.R.C. Post, P. Ashburn and G.R. Wolstenholme, 'Polysilicon emitters for bipolar transistors: a review and re-evaluation of theory and experiment', *IEEE Trans. Electron. Devices*, **39**, 1717 (1992).

[13] Z. Yu, B. Ricco and R.W. Dutton, 'A comprehensive analytical and numerical model of polysilicon emitter contacts in bipolar transistors', *IEEE Trans. Electron. Devices*, **31**, 773 (1984).

[14] P. Ashburn, *Design and Realization of Bipolar Transistors*, Wiley (1988).

[15] P. Ashburn, D.J. Roulston and C.R. Selvakumar, 'Comparison of experimental and computed results on arsenic and phosphorus doped polysilicon emitter bipolar transistors', *IEEE Trans. Electron. Devices*, **34**, 1346 (1987).

[16] N.E. Moiseiwitsch and P. Ashburn, 'The benefits of fluorine in *pnp* polysilicon emitter bipolar transistors', *IEEE Trans. Electron. Devices*, **41**, 1249 (1994).
[17] G.L. Patton, J.C. Bravman and J.D. Plummer, 'Physics, technology, and modeling of polysilicon emitter contacts for VLSI bipolar transistors', *IEEE Trans. Electron. Devices*, **33**, 1754 (1986).
[18] A. Neugroschel, M. Arienzo, Y. Komem and R.D. Isaac, 'Experimental study of the minority carrier transport at the polysilicon-monosilicon interface', *IEEE Trans. Electron. Devices*, **32**, 807 (1985).
[19] C.D. Marsh, N.E. Moiseiwitsch, G.R. Booker and P. Ashburn, 'Behaviour and effects of fluorine in annealed *n*+ polycrystalline silicon layers on silicon wafers', *Jnl Appl. Phys.* **87**, 7567 (2000).
[20] J.D. Williams and P. Ashburn, 'Epitaxial regrowth of *n*+ and *p*+ polycrystalline silicon layers given single and double diffusions', *Jnl Appl. Phys.* **72**, 3169 (1992).
[21] J.S. Hamel, D.J. Roulston and C.R. Selvakumar, 'Two dimensional analysis of emitter resistance in the presence of interfacial oxide break-up in polysilicon emitter bipolar transistors', *IEEE Trans. Electron. Devices*, **39**, 2139 (1992).
[22] M. Hendriks, 'Interface engineering in silicon semiconductor processing using a cluster tool', *Appl. Surface Sci.* **70/71**, 619 (1993).
[23] S. Decoutere, A. Cuthbertson, R. Wilhelm, W. Vandervorst and L. Deferm, 'Engineering of the polysilicon emitter interfacial layer using low temperature thermal re-oxidation in an LPCVD cluster tool', *Proc. ESSDERC*, 429 (1995).
[24] F.S. Becker, H. Oppolzer, I. Weitzel, H. Eichermuller and H. Schaber, 'Low resistance polycrystalline silicon by boron or arsenic implantation and thermal crystallization of amorphously deposited films', *Jnl Appl. Phys.*, **56**, 1233 (1984).
[25] Y. Wada and S. Mishimatsu, 'Grain growth mechanism of heavily doped phosphorus-implanted polycrystalline silicon', *Jnl Electrochem. Soc.*, **125**, 1499 (1978).
[26] T. Hashimoto, T. Kumauchi, T. Jinbo, K. Watanabe, E. Toshida, H. Miura, T. Shiba and Y. Tamaki, 'Interface controlled IDP process technology for 0.3 μm high speed bipolar polysilicon emitter transistors', *Proc. BCTM*, 181 (1996).
[27] I.R.C. Post and P. Ashburn, 'Investigation of boron diffusion in polysilicon and its application to the design of *pnp* polysilicon emitter bipolar transistors with shallow emitter junctions', *IEEE Trans. Electron. Devices*, **38**, 2442 (1991).
[28] I.R.C. Post, P. Ashburn and A. Nouailhat, 'A heterojunction tunnelling model for *pnp* and *npn* polysilicon emitter bipolar transistors', *Electron. Lett.*, **28**, 2276 (1992).

[29] I.R.C. Post, P. Ashburn and A. Nouailhat, 'An investigation of the inconsistency in barrier heights for *pnp* and *npn* polysilicon emitter bipolar transistors using a new tunneling model', *Japanese Jnl. Appl. Phys.*, 1275 (1994).

7

Properties and Growth of Silicon-Germanium

7.1 INTRODUCTION

Silicon and germanium are completely miscible over the full range of compositions and hence can be combined to form $Si_{1-x}Ge_x$ alloys with the germanium content x ranging from 0 to 1 (0–100%). The property of $Si_{1-x}Ge_x$ that is of interest for bipolar transistors is the bandgap, which is smaller than that of silicon and controllable by varying the germanium content. Bandgap engineering concepts that were previously only possible in compound semiconductor technologies, have now become viable in silicon technology. These concepts have introduced a new degree of freedom in the design of the base that have allowed the base doping to be increased and the basewidth to be reduced, while at the same time maintaining a reasonable value of gain. In this way, much higher values of f_T and f_{max} have been achieved.

While $Si_{1-x}Ge_x$ alloys have been researched since the late 1950s [1], it is only in the past ten years or so that these layers have been applied to bipolar technology [2,3]. This has been made possible by the development of new growth techniques, such as Molecular Beam Epitaxy (MBE), Low Pressure Chemical Vapour Deposition (LPCVD) and Ultra-High Vacuum Chemical Vapour Deposition (UHV-CVD). The key feature of these techniques that has led to the development of SiGe Heterojunction Bipolar Transistors (HBTs), is the growth of epitaxial layers at low temperatures (300–800°C). This allows very narrow bases to be grown with sharp doping profiles and also $Si_{1-x}Ge_x$

SiGe Heterojunction Bipolar Transistors Peter Ashburn

ISBN: 0-470-84838-3

layers to be grown on a silicon substrate, even though there is a lattice mismatch between silicon and germanium of 4.2%. In this chapter, the materials properties of silicon-germanium will first be outlined, and then the methods used for growing $Si_{1-x}Ge_x$ layers will be described. The theory of SiGe HBTs will be described in Chapter 8 and SiGe HBT technology will be covered in Chapter 10.

7.2 MATERIALS PROPERTIES OF SILICON-GERMANIUM

7.2.1 Pseudomorphic Silicon-Germanium

$Si_{1-x}Ge_x$ has a diamond-like lattice structure and the lattice constant is given by Vegard's rule:

$$a_{Si_{1-x}Ge_x} = a_{Si} + x(a_{Ge} - a_{Si}) \tag{7.1}$$

where x is the germanium fraction and a is the lattice constant. The lattice constant of silicon a_{Si} is 0.543 nm, the lattice constant of germanium a_{Ge} is 0.566 nm and the lattice mismatch is 4.2%.

When a $Si_{1-x}Ge_x$ layer is grown on a silicon substrate, the lattice mismatch at the interface between the $Si_{1-x}Ge_x$ and the silicon has to be accommodated. This can either be done by compression of the $Si_{1-x}Ge_x$ layer so that it fits to the silicon lattice or by the creation of misfit dislocations at the interface. These two possibilities are illustrated schematically in Figure 7.1. In the former case, the $Si_{1-x}Ge_x$ layer adopts the silicon lattice spacing in the plane of the growth and hence the normally cubic $Si_{1-x}Ge_x$ crystal is distorted. When $Si_{1-x}Ge_x$ growth

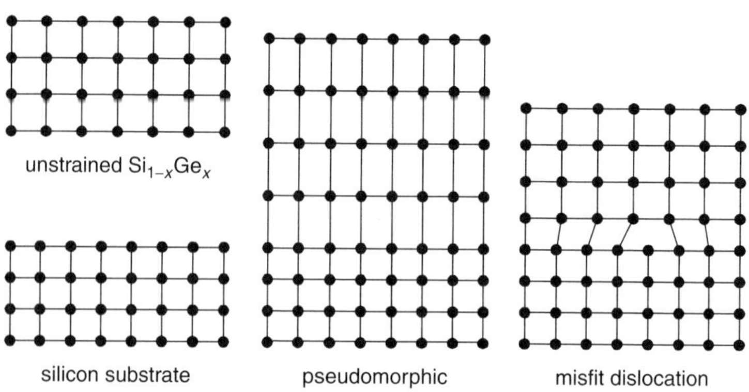

Figure 7.1 Schematic illustration of pseudomorphic $Si_{1-x}Ge_x$ growth and misfit dislocation formation

occurs in this way, the $Si_{1-x}Ge_x$ layer is under compressive strain and the layer is described as *pseudomorphic*. In the latter case, the $Si_{1-x}Ge_x$ layer is unstrained, or relaxed, and the lattice mismatch at the interface is accommodated by the formation of misfit dislocations. These misfit dislocations generally lie in the plane of the interface, as shown in Figure 7.1, but dislocations can also thread vertically through the $Si_{1-x}Ge_x$ layer.

7.2.2 Critical Thickness

As might be expected, there is a maximum thickness of $Si_{1-x}Ge_x$ that can be grown before relaxation of the strain occurs through the formation of misfit dislocations. This is known as the critical thickness of the $Si_{1-x}Ge_x$ layer, and depends strongly on the germanium content, as shown in Figure 7.2. The original calculations of critical layer thickness were made by Matthews and Blakeslee [4,5] on the basis of the mechanical equilibrium of an existing threading dislocation. However, measurements of dislocation densities in $Si_{1-x}Ge_x$ showed, in many cases, no evidence of misfit dislocations for $Si_{1-x}Ge_x$ layers considerably thicker than the Matthews–Blakeslee limit. These results were explained by People and Bean [6] who calculated the critical thickness

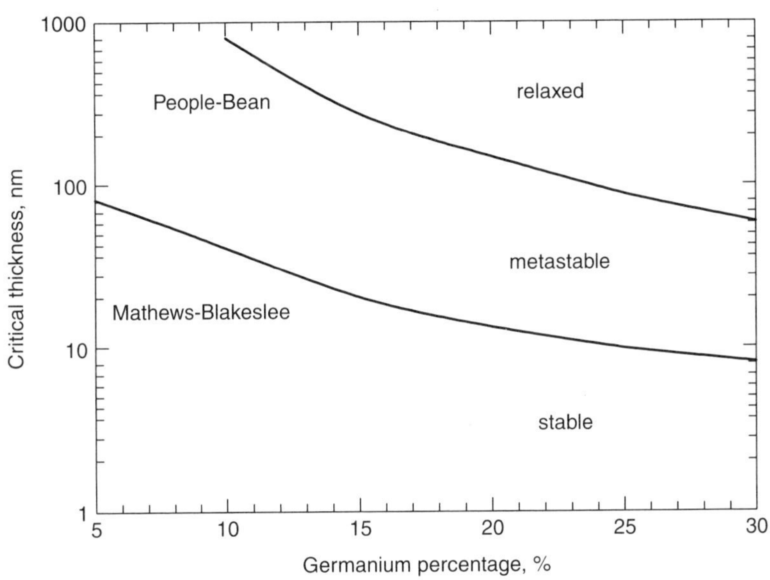

Figure 7.2 Critical $Si_{1-x}Ge_x$ thickness as a function of germanium percentage (reprinted with permission from [2])

on the assumption that misfit dislocation generation was determined solely by energy balance. The discrepancy between these two types of calculation can be explained by the observation that strain relaxation in $Si_{1-x}Ge_x$ layers occurs gradually. Layers above the People–Bean curve can be considered to be completely relaxed, whereas layers below the Matthews–Blakeslee curve can be considered to be fully strained. These fully strained layers are termed *stable* and will not relax during any subsequent high-temperature processing. Layers lying between the two curves are termed *metastable*; these layers may be free of dislocations after growth, but are susceptible to relaxation during later high-temperature processing.

In practice, a number of additional factors influence the critical thickness of a $Si_{1-x}Ge_x$ layer. Of particular importance to $Si_{1-x}Ge_x$ HBTs is the effect of a silicon cap layer, which has been shown to increase the critical thickness of the underlying $Si_{1-x}Ge_x$ layer. Figure 7.3 shows a comparison of the calculated critical thickness as a function of germanium percentage for stable $Si_{1-x}Ge_x$ layers with and without a silicon cap. It can be seen that the critical thickness is more than doubled by the presence of the silicon cap.

The presence of misfit dislocations in $Si_{1-x}Ge_x$ HBTs is highly undesirable, since they create generation/recombination centres, which

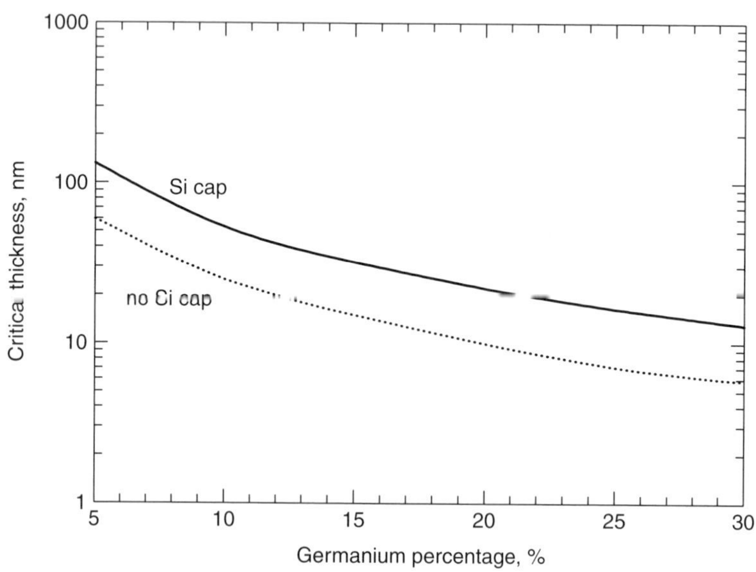

Figure 7.3 Critical thickness as a function of germanium percentage for stable $Si_{1-x}Ge_x$ layers with and without a silicon cap (reprinted with permission from [7])

degrade the low-current gain, as discussed in Section 4.2.4. Threading dislocations also highly undesirable, as they can lead to the formation of emitter/collector pipes. When designing the base of a $Si_{1-x}Ge_x$ HBT, it is important that the $Si_{1-x}Ge_x$ thickness is chosen to give a stable base, so that dislocation formation is avoided. A base thickness below the silicon cap curve in Figure 7.3 will ensure a stable base, which will withstand ion implantation and high-temperature annealing without encountering problems of relaxation and misfit dislocation generation.

Considerable research has been done on the oxidation of $Si_{1-x}Ge_x$ [8,9], and it has been found that the germanium in the $Si_{1-x}Ge_x$ layer does not oxidize, but piles up at the oxide/$Si_{1-x}Ge_x$ interface. This pile-up of germanium makes it difficult to achieve low values of interface state density in oxidized $Si_{1-x}Ge_x$ layers. It is therefore advisable to avoid direct oxidation of the $Si_{1-x}Ge_x$ layer in $Si_{1-x}Ge_x$ HBT technologies. This is generally easy to achieve if the $Si_{1-x}Ge_x$ base is buried below a silicon cap layer.

7.2.3 Band Structure of Silicon-Germanium

$Si_{1-x}Ge_x$ alloys have a smaller bandgap than silicon partly because of the larger lattice constant and partly because of the strain. Figure 7.4

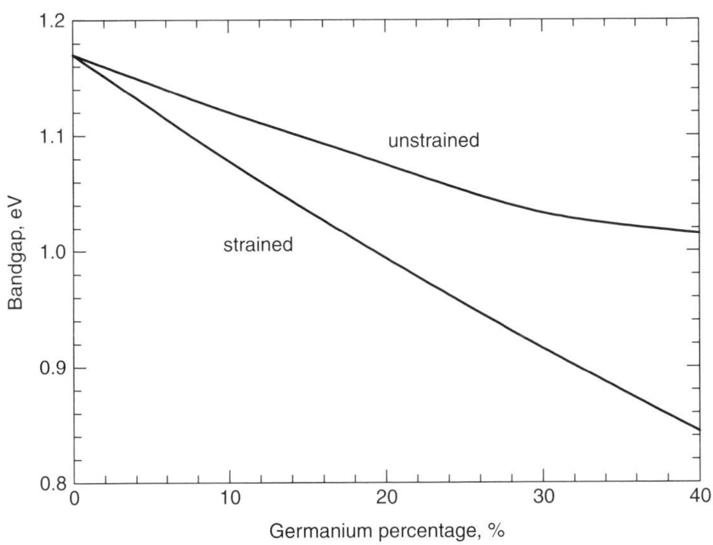

Figure 7.4 Bandgap as a function of germanium percentage for strained [10] and unstrained [1] $Si_{1-x}Ge_x$ (reprinted with permission from [2])

shows the variation of bandgap with germanium percentage for strained and unstrained $Si_{1-x}Ge_x$. It can be seen that the strain has a dramatic effect on the bandgap of $Si_{1-x}Ge_x$. For 10% germanium, the reduction in bandgap compared with silicon is 92 meV for strained $Si_{1-x}Ge_x$, compared with 50 meV for unstrained $Si_{1-x}Ge_x$. The variation of bandgap with germanium content for strained $Si_{1-x}Ge_x$ can be described by the following empirical equation:

$$E_G(x) = 1.17 - 0.96x + 0.43x^2 - 0.17x^3 \tag{7.2}$$

The band alignment for compressively strained $Si_{1-x}Ge_x$ on unstrained silicon is illustrated schematically in Figure 7.5. This band alignment is referred to as *type I*, and the majority of the band offset at the heterojunction interface occurs in the valence band, with only a small offset in the conduction band. Different band alignments can be obtained by engineering the strain in the substrate and the grown layer in different ways. For example, *type II* band alignments can be obtained by growing tensilely strained silicon on top of unstrained $Si_{1-x}Ge_x$. This arrangement gives large conduction and valence band offsets and is used in strained silicon heterojunction MOSFETs.

Figure 7.6 shows the variation of valence band offset ΔE_V, conduction band offset ΔE_C, and bandgap difference ΔE_G, with germanium content. It can be seen that the majority of the band offset occurs in the valence band. For example for 10% germanium, the valence band offset is 0.073 eV, compared with 0.019 eV for the conduction band offset. The conduction band offset can therefore be neglected for most practical purposes.

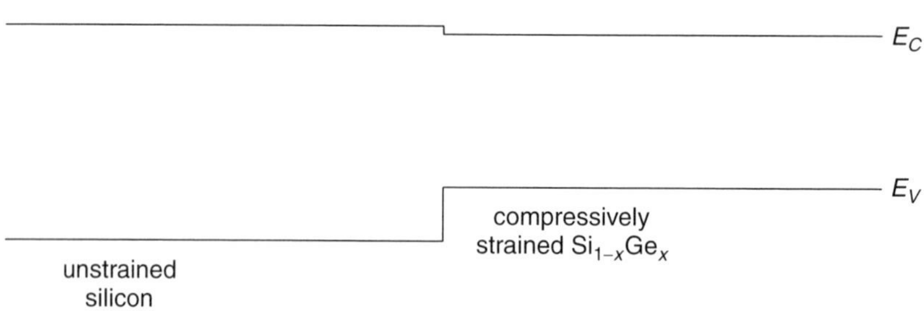

Figure 7.5 Illustration of the band alignment obtained for a compressively strained $Si_{1-x}Ge_x$ layer grown on an unstrained silicon substrate

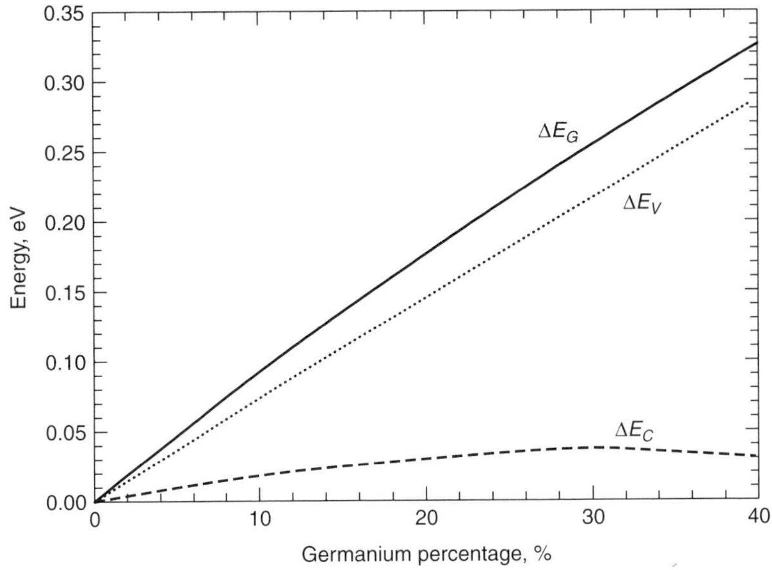

Figure 7.6 Valence band offset ΔE_V, conduction band offset ΔE_C and bandgap difference ΔE_G as a function of germanium percentage for strained $Si_{1-x}Ge_x$ grown on unstrained silicon (reprinted with permission from [11])

7.3 PHYSICAL PROPERTIES OF SILICON-GERMANIUM

7.3.1 Dielectric Constant

The dielectric constant of $Si_{1-x}Ge_x$ can be obtained by linear interpolation between the known values for silicon and germanium [11] using the following equation:

$$\varepsilon(x) = 11.9(1 + 0.35x) \quad (7.3)$$

7.3.2 Density of States

Although, the density of states in the conduction band in $Si_{1-x}Ge_x$ is generally assumed to be the same as that in silicon, there is some evidence in the literature to suggest that the density of states in the valence band is considerably smaller. Manku and Nathan [12,13] have calculated an $E - k$ diagram for strained $Si_{1-x}Ge_x$ and shown that the density of states hole mass is significantly lower, by a factor of approximately three at 30% germanium. There is some experimental evidence to support this calculation. For example, freeze-out of holes in p-type $Si_{1-x}Ge_x$ has been reported to occur at higher temperatures than in p-type silicon [14] and enhancements in the majority carrier hole mobility have been reported for p-type $Si_{1-x}Ge_x$ [15].

Using the calculated values of hole density of states of Manku and Nathan [12,13], the hole concentration can be calculated as a function of Fermi level position. These results are shown in Figure 7.7 for $Si_{1-x}Ge_x$ with four different germanium contents. It can be seen that the Fermi level moves deeper in the valence band as the germanium concentration increases. Figure 7.8 shows the ratio of the calculated density of states in the valence band for $Si_{1-x}Ge_x$ to that for silicon as a function of germanium content. It is clear that the density of states in the valence band for $Si_{1-x}Ge_x$ is significantly lower than that for silicon at germanium contents of practical interest.

7.3.3 Apparent Bandgap Narrowing

In practical $Si_{1-x}Ge_x$ HBTs, there will be two sources of bandgap narrowing in the base, one due to the strained $Si_{1-x}Ge_x$ and one due to the heavy doping as discussed in Section 3.3. Unfortunately very little work

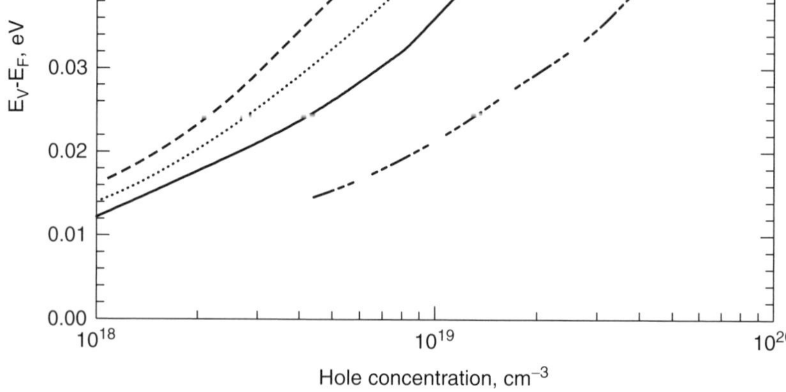

Figure 7.7 Fermi level position as a function of hole concentration for $Si_{1-x}Ge_x$ with four different germanium concentrations (reprinted with permission from [2])

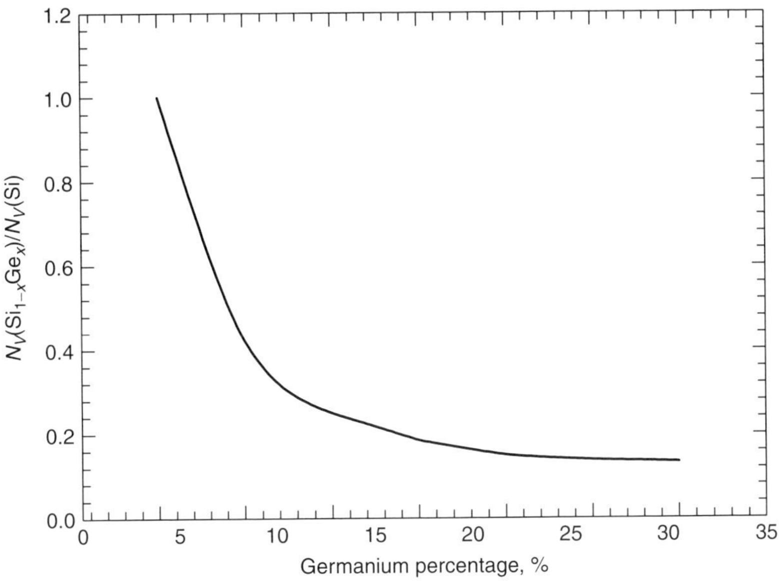

Figure 7.8 Ratio of density of states in the valence band for $Si_{1-x}Ge_x$ to that for Si as a function of germanium percentage (reprinted with permission from [14])

has been published on this subject and hence a consensus has not fully emerged. In this section, the theoretical approach of [11] is followed, which has been shown to be in reasonable agreement with experiment. Figure 7.9 shows the apparent bandgap narrowing in $Si_{1-x}Ge_x$ as a function of acceptor concentration for three values of germanium content. At low acceptor concentrations, the apparent bandgap narrowing in $Si_{1-x}Ge_x$ is slightly higher than that in silicon, but at acceptor concentrations in the range $1-2 \times 10^{19}$ cm^{-3} the apparent bandgap narrowing is approximately the same. This latter doping range is the base doping range that is of practical interest for $Si_{1-x}Ge_x$ HBTs.

7.3.4 Minority Carrier Hole Mobility

Unfortunately very few measurements of minority carrier hole mobility have been made in $Si_{1-x}Ge_x$. Poortmans [14] inferred values of minority carrier hole mobility from measurements on $Si_{1-x}Ge_x$ HBTs and found an enhancement in mobility compared with silicon by a factor of 1.2–1.4 for base doping concentrations in the range $5 \times 10^{18}-5 \times 10^{19}$ cm^{-3}. Given the scarcity of measured data on minority carrier mobility and density of states in $Si_{1-x}Ge_x$, the most reliable way of calculating the

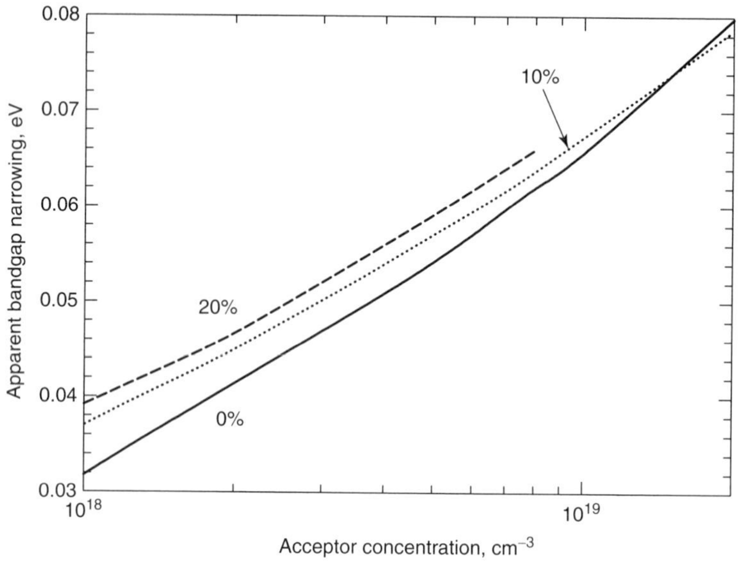

Figure 7.9 Apparent bandgap narrowing as a function of acceptor concentration for $Si_{1-x}Ge_x$ with three different germanium concentrations (reprinted with permission from [11])

expected gain improvement in a $Si_{1-x}Ge_x$ HBT is to use data directly obtained from measurements on $Si_{1-x}Ge_x$ HBTs. As will be shown in Section 8.3, the gain enhancement in a $Si_{1-x}Ge_x$ HBT is determined by the ratio of the product $N_C N_V D_{nb}$ in $Si_{1-x}Ge_x$ and Si, together with the bandgap narrowing due to the strained $Si_{1-x}Ge_x$. Figure 7.10 shows a graph of this $N_C N_V D_{nb}$ ratio as a function of acceptor concentration for three values of germanium content. It can be seen that for germanium contents of practical interest, in the range 11–16%, this ratio has a value of around 0.25.

7.4 BASIC EPITAXY THEORY

In this section a simple model will be described that explains the essential features of epitaxial $Si_{1-x}Ge_x$ growth. The model is given in Figure 7.11 and shows the distribution of the reactant species in the gas. The concentration of the reactant in the bulk of the gas is given by C_G and the concentration at the surface of the film is represented by C_S. The epitaxial growth occurs by the transport of the reactant species from the bulk of the gas to the surface of the film and the chemical reaction of the reactant with the film at the surface.

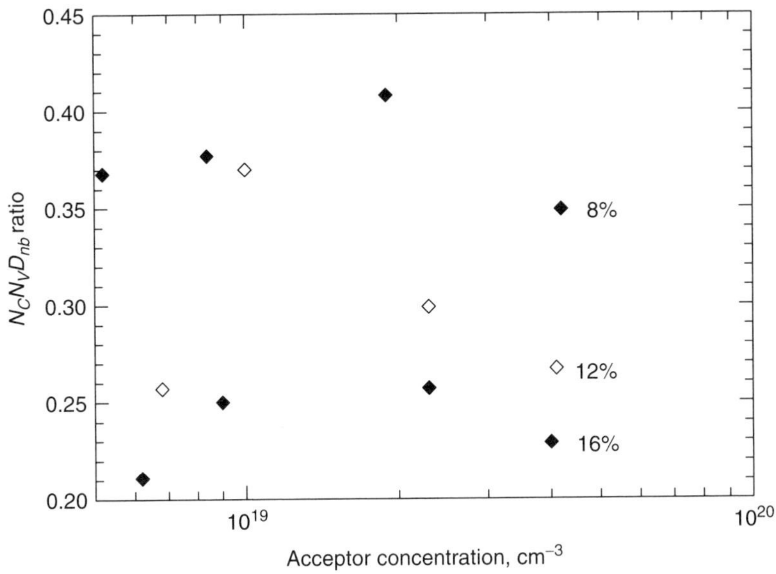

Figure 7.10 Measured values of the ratio of $N_C N_V D_{nb}$ in $Si_{1-x}Ge_x$ to that in Si as a function of acceptor concentration (reprinted with permission from [14])

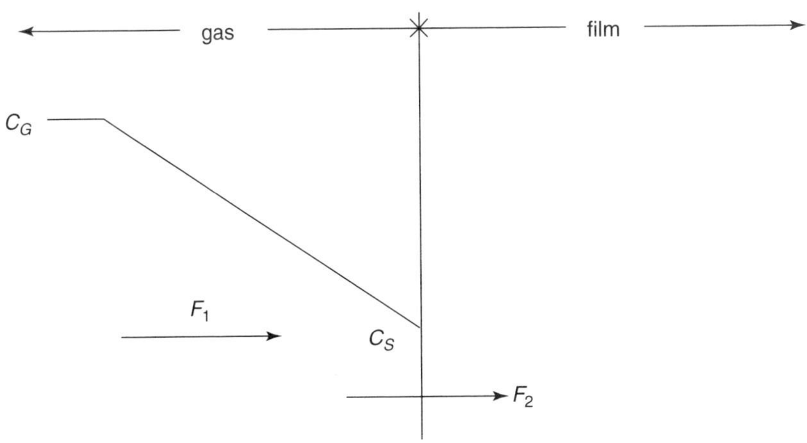

Figure 7.11 Simple model of the epitaxial growth process

The flux of reactant from the bulk of the gas to the film surface can be represented by the following simple linear distribution:

$$F_1 = h_G(C_G - C_S) \tag{7.4}$$

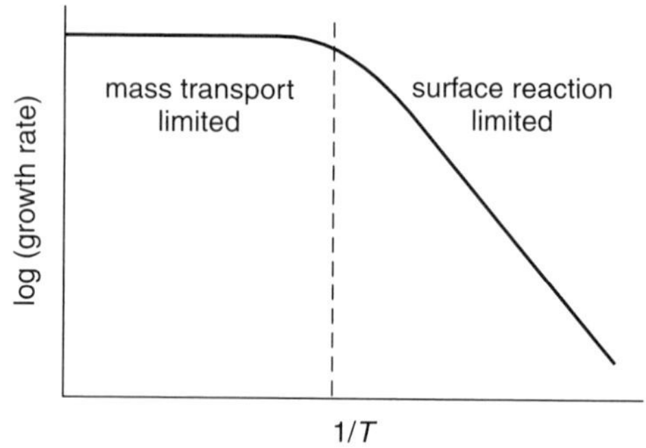

Figure 7.12 Schematic illustration of the temperature dependence of the growth rate, showing the mass transport limited and surface reaction limited regimes

where h_G is the gas phase mass transport coefficient. The flux consumed by the chemical reaction at the surface of the growing film can be approximated by:

$$F_2 = k_S C_S \tag{7.5}$$

where k_S is the surface reaction rate constant. In steady state, $F_1 = F_2$, giving:

$$C_S = \frac{h_G}{h_G + k_S} C_G \tag{7.6}$$

The growth rate of the film is given by:

$$G_R = \frac{F_2}{N_F} = \frac{k_S h_G}{h_G + k_S} \frac{C_G}{N_F} \tag{7.7}$$

where N_F is the number of silicon atoms incorporated into a unit volume of the film. In general, there may be more than one species in the gas, in which case C_G in equation (7.7) is replaced by $\gamma_M C_T$, where γ_M is the mole fraction of the reactant species and C_T is the total number of molecules per unit volume in the gas.

These equations predict the basic behaviour seen in the growth of epitaxial films. Equation (7.6) shows that the surface concentration approaches zero if $k_S \gg h_G$. This growth regime is referred to as the mass transport limited regime. Similarly the surface concentration approaches C_G when $h_G \gg k_S$. This growth regime is referred to as the surface reaction limited regime. The growth rates for these two limiting cases are given by:

$$\text{Surface reaction controlled } G_R = \frac{C_G}{N_F} k_S \quad (7.8)$$

$$\text{Mass transport controlled } G_R = \frac{C_G}{N_F} h_G \quad (7.9)$$

Chemical reaction rates normally vary exponentially with temperature, so in the surface reaction limited regime the growth rate varies strongly with temperature. Alternatively, the mass transport coefficient h_G is relatively insensitive to temperature, so in the mass transport limited regime, the growth rate varies very little with temperature. These two limiting growth regimes are illustrated schematically in Figure 7.12.

7.4.1 Boundary Layer Model

To estimate the mass transport coefficient, a model for the gas flow adjacent to the wafer is needed. In the bulk of the gas, the flow of gas will be uniform with a velocity V_G, whereas adjacent to the substrate the flow of gas must be zero. The boundary layer adjacent to the substrate will therefore have the form shown in Figure 7.13, where the thickness of the boundary layer t_b varies with horizontal distance x, and the gas velocity in the boundary layer v_G varies with vertical distance y.

The force resisting the flow of the gas adjacent to the substrate is the friction F, which is given by:

$$F = v_{isc} \frac{\partial v_G}{\partial y} \quad (7.10)$$

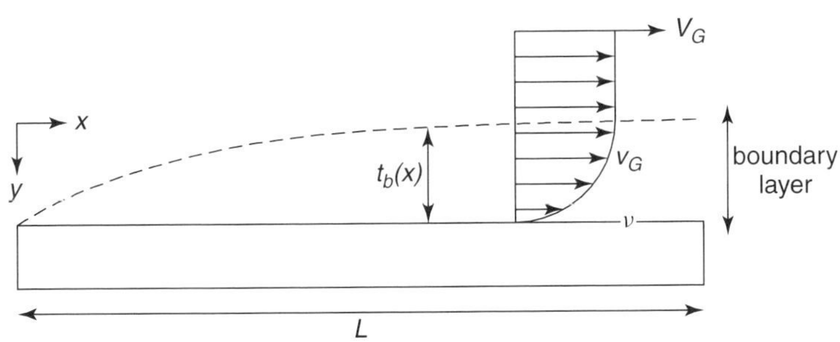

Figure 7.13 Boundary layer model for epitaxial growth

where v_{isc} is the viscosity of the gas. Using Newton's second law, we have:

$$F = ma = m\frac{dv_G}{dt} = m\frac{dv_G}{dx}\frac{dx}{dt} = m\frac{dv_G}{dx}v_G \tag{7.11}$$

where m is the mass of an element of gas, which is given by $\rho_G t_b(x)dx$, where ρ_G is the density of the gas. Combining equations (7.10) and (7.11) and replacing the differentials by their respective differences gives:

$$v_{isc}\frac{V_G}{t_b(x)} = \rho_G t_b(x) V_G \frac{V_G}{x} \tag{7.12}$$

This equation can be rearranged to give the variation of the thickness of the boundary layer with distance x:

$$t_b(x) = \sqrt{\frac{v_{isc}x}{\rho_G V_G}} \tag{7.13}$$

The average boundary layer thickness can be obtained by integrating along the length L of the substrate to give:

$$t_b = \frac{2}{3}L\sqrt{\frac{v_{isc}L}{\rho_G V_G}} = \frac{2}{3}\frac{L}{\sqrt{R_e}} \tag{7.14}$$

where R_e is the Reynolds number for the gas. The Reynolds number is an extremely important dimensionless quantity in fluid dynamics and it determines the characteristics of the gas flow. For small values of Reynolds number (<2000) the flow is laminar, whereas for large values of Reynolds number the flow is turbulent. From equation (7.14), it can be seen that the Reynolds number is given by:

$$R_e = \frac{\rho_G V_G L}{v_{isc}} \tag{7.15}$$

In the boundary region model, the mass transport coefficient h_G is given by the ratio of the diffusion coefficient of the reactant species D_G divided by the average boundary layer thickness:

$$h_G = \frac{D_G}{t_b} = \frac{3}{2}\frac{D_G}{L}\sqrt{R_e} \tag{7.16}$$

This equation, together with equation (7.9), shows that in the mass transport limited regime the growth rate is proportional to the square root of the gas velocity V_G. In real epitaxial reactors, the boundary layers at the reactor walls also have to be taken into account. These considerations indicate that in practice the geometry of the susceptor on which the wafers sit and the geometry of the reactor will influence the growth rate.

7.4.2 Growth Modes

When the growth is considered at the atomistic level, other factors can be seen to influence the epitaxial growth. When the reactant species arrive at the surface of the film, adsorption of the molecules at the surface occurs. These adatoms are usually weakly bonded to the film surface and thus are able to diffuse across the surface to find a nucleation site. These nucleation sites could be steps in the growth plane, islands of growth or impurity atoms. The final stage is desorption of volatile reaction biproducts from the film surface. The structure of the resulting film is a strong function of the adsorption and surface diffusion rates. If the surface diffusion is slow compared with the rate of arrival of the reactant species, amorphous films are produced. Conversely, if the surface diffusion is fast relative to the incoming flux, single-crystal growth occurs. Amorphous films are produced at low temperatures, single-crystal films at high temperatures, and polycrystalline films at intermediate temperatures.

Figure 7.14 illustrates the nucleation and growth modes that have been observed in the practical growth of epitaxial films. Layer-by-layer growth occurs when atoms are equally or less strongly bonded to each other than to the substrate, and is the usual mode observed in the growth of epitaxial $Si_{1-x}Ge_x$ films. The presence of contamination, such as carbon or oxygen, on the substrate surface prior to growth can disrupt layer-by-layer growth and lead to 3D island growth. 3D island growth occurs when arriving atoms are more strongly bonded to each other than to the substrate, and it leads to the nucleation of small clusters on the substrate surface. This is the case for Si or $Si_{1-x}Ge_x$ growth on silicon dioxide. Layer-plus-island, or Stranski-Krastanov, growth is a combination of the previous two growth mechanisms. It usually occurs after the layer-by-layer growth of one or two monolayers of the film, after which layer-by-layer growth becomes unfavourable and islands form on top of the initial layers. Although 3D island growth is undesirable for $Si_{1-x}Ge_x$ HBTs, it is widely used for the growth of quantum dots.

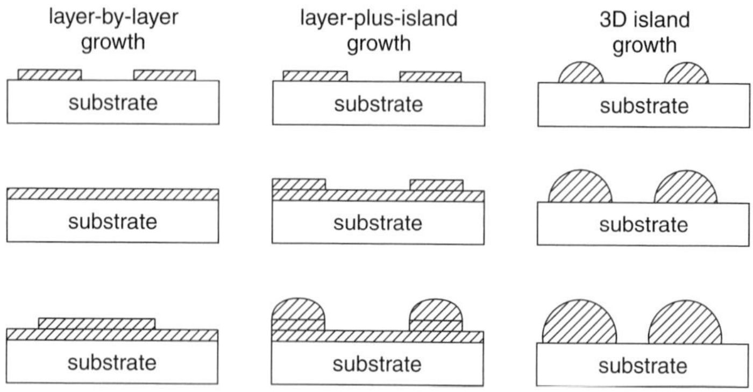

Figure 7.14 Growth modes obtainable in epitaxial films

7.5 LOW-TEMPERATURE EPITAXY

Over the past ten years and more there have been rapid developments in techniques for the growth of $Si_{1-x}Ge_x$ epitaxial layers at low temperatures. This has been made possible by a number of changes in the design of epitaxy equipment and by improvements to growth processes. There are two main prerequisites for the growth of epitaxial layers at low temperature: establishment of a clean surface prior to growth, and growth in an ultra-clean environment.

The removal of oxygen and carbon is the main problem in establishing a clean surface prior to growth. A clean silicon surface is highly reactive and oxidizes in air even at room temperature. The secret of low-temperature epitaxial growth is therefore the removal of this native oxide layer and the maintenance of a clean surface until epitaxy can begin. Two alternative approaches to pre-epitaxy surface cleaning have been developed, as described below.

7.5.1 In situ Hydrogen Bake

The concept that underlies this surface clean is the controlled growth of a thin surface oxide layer, followed by its removal in the epitaxy reactor using a hydrogen bake. The controlled growth of the surface oxide layer is generally achieved using an RCA clean or a variant. The RCA clean is a two-stage clean that is widely used in the silicon industry to clean wafers prior to oxidation or anneal. The first stage, RCA1, comprises a 10 minute soak in a hot solution of hydrogen peroxide, ammonium hydroxide and water, in the volume ratio of 1:1:5. The purpose of RCA1 is to remove carbon contamination from organic chemicals, such as photoresist, used in lithography. RCA1 can sometimes lead to surface roughening, which can be eliminated by reducing the ammonia content [16]. Depending on the cleanliness of the chemicals used, RCA1 can leave some trace metal contamination. This is removed in RCA2, which comprises a ten minute soak in a hot solution of hydrogen peroxide, hydrochloric acid and water, in the volume ratio 1:1:5. The RCA clean produces a hydrophilic surface with an oxide thickness of around 1.5 nm.

The oxide created by the RCA clean is removed in the reactor using an in situ bake in hydrogen for around 15 minutes at a temperature in the range 900–950°C. The temperature required to remove the native oxide depends on the thickness of the oxide, which is determined by the severity of the surface clean. Special pre-epitaxy cleans have been developed that allow surface oxide removal at lower temperatures. For example, the Shiraki clean [17] enables the surface oxide to be removed using a 15 minute hydrogen bake at 750°C. The Shiraki clean uses essentially the same chemicals as the RCA clean, but in the last step the amount of hydrochloric acid in the solution is increased to give a volume ratio of 1:3:1 of H_2O_2:HCl:H_2O. The increased amount of HCl is reported to create a surface oxide that is more easily removed due to the presence of SiO_xH_{y-} compounds in the oxide that protect against carbon contamination.

7.5.2 Hydrogen Passivation

An alternative approach to pre-epitaxy cleaning is to create an oxide-free surface using an ex situ clean and then move quickly to epitaxial growth before the native oxide can grow. The aim of the ex situ clean is to produce a surface that is passivated by hydrogen atoms bonded to dangling bonds from silicon atoms on the surface. When the wafers are transferred in the epitaxy reactor, the hydrogen can be released from the surface of the silicon very quickly using a low-temperature bake or even in the early stages of epitaxy without any bake. Meyerson [18] has reported that hydrogen desorbs at 600°C at a rate of a few monolayers per second, so the hydrogen passivation approach allows epitaxial layers to be grown at low temperatures without the need for a high-temperature bake.

The ex situ cleaning cycle begins with an RCA clean to remove surface contamination and is followed by a 30 second dip etch in dilute hydrofluoric acid (typically 2–5%). The wafers are then blow-dried or spin-dried to maintain the hydrogen passivated surface. The hydrogen passivation is stable for typically 30 minutes after completion of the ex situ cleaning [19]. It is important that the wafers should not be given lengthy rinses in water after removal from the hydrofluoric acid because this can destroy the hydrogen passivation and lead to the introduction of contaminants such as oxygen and carbon.

7.5.3 Ultra-clean Epitaxy Systems

Having produced a clean, hydrogen passivated silicon surface, it is clearly important to maintain the state of this surface in the epitaxy system. This necessitates the use of low-pressure epitaxy systems if epitaxial growth at low temperatures is required. Figure 7.15 summarizes the partial pressures of oxygen and water vapour that need to be achieved in an epitaxy system if an oxide-free surface is to be maintained at a given temperature [20,21]. This figure shows that epitaxial growth at low temperature requires low partial pressures of oxygen and water vapour, which of course can be achieved by reducing the pressure in the epitaxy system. Research [22] has shown that a pressure below 30 Torr is needed to achieve silicon epitaxial growth below 900°C.

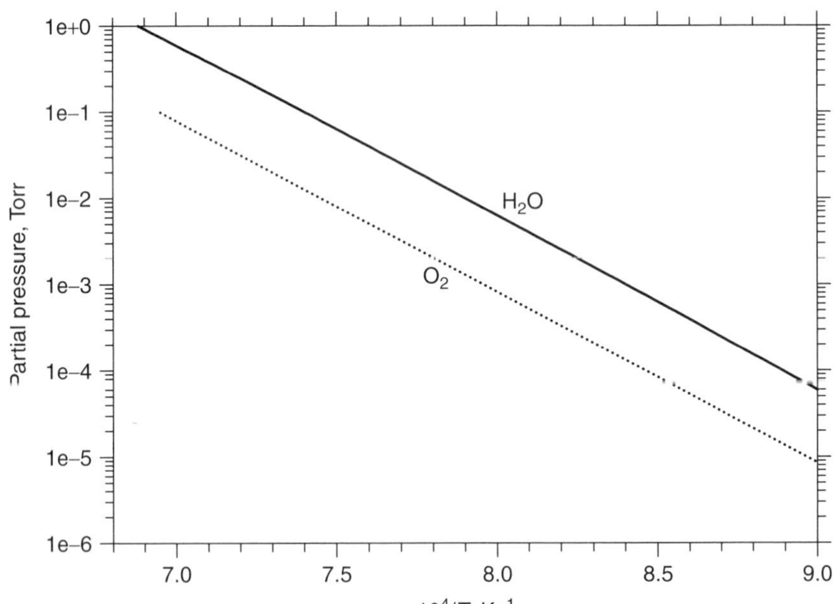

Figure 7.15 Conditions for oxide formation in an epitaxy system. Note that 1 atm = 1.113 bar = 760 Torr = 1.113×10^5 Pa (reprinted with permission from the Electrochemical Society [20,21])

Figure 7.15 implies that the following requirements need to be met in an epitaxy system to achieve epitaxial growth at low temperature:

- purification of all inlet gases;
- use of a loadlock to prevent contamination during wafer loading;
- use of hydrocarbon-free pumping systems (e.g. turbo-molecular or cryo-pumping);
- operation at reduced pressure to reduce the partial pressure of oxygen and water vapour.

7.6 COMPARISON OF SILICON AND SILICON-GERMANIUM EPITAXY

Epitaxial growth of silicon can be achieved over a wide range of temperatures using Low-Pressure Chemical Vapour Deposition (LPCVD) [22]. Figure 7.16 shows the growth rate as a function of reciprocal temperature for three different growth gases. For temperatures above 800–850°C, the growth rate varies very little with temperature, indicating that growth is mass transport limited. For temperatures below 800°C, the growth rate varies strongly with temperature, indicating that

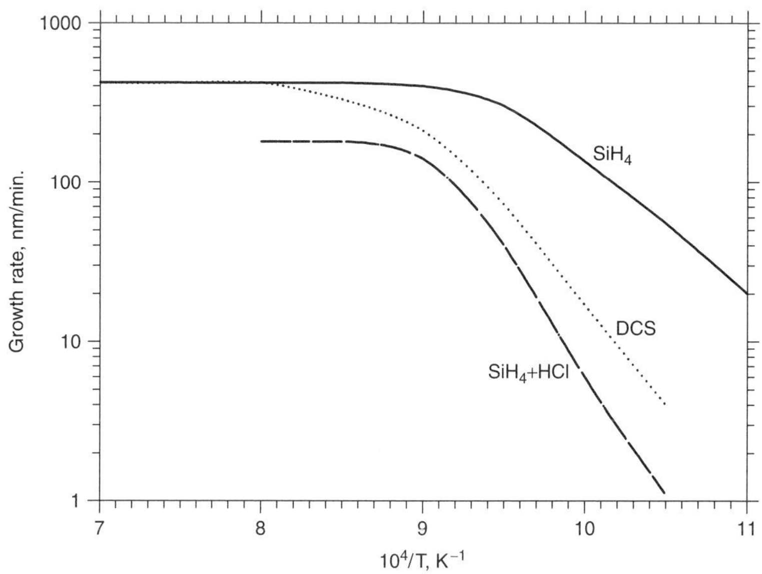

Figure 7.16 Silicon growth rate as a function of reciprocal temperature for three different growth gases: 40 sccm of silane (SiH_4), 80 sccm of dichlorosilane (SiH_2Cl_2) and 20 sccm of SiH_4 with 2 sccm of HCl. The hydrogen flow was 2 slm (reprinted with permission from American Institute of Physics [22])

growth is surface reaction limited. These results indicate that growth at low temperatures requires growth in the surface reaction limited regime whereas growth at high temperatures is done in the mass transport limited regime. For growth using dichlorosilane and silane plus HCl, reasonable growth rates can be obtained down to a temperature of around 650°C, while for silane reasonable growth rates can be obtained at even lower temperatures. With growth using LPCVD, the process pressure is typically around 1 Torr. Dopants are incorporated using diborane B_2H_6, arsine AsH_3 and phosphine PH_3.

Growth at even lower temperatures can be achieved using Ultra-High Vacuum, Chemical Vapour Deposition (UHV-CVD) [23,24]. With UHV-CVD, the system is designed to deliver a very low base pressure of around 10^{-9} Torr to reduce the partial pressures of oxygen and water vapour in the system. This allows a lower pressure regime to be used for the growth of around 10^{-3} Torr. Growth rates of 0.5–10 nm/min can be achieved at growth temperatures down to 550°C.

The growth of $Si_{1-x}Ge_x$ epitaxial layers is done using the same equipment and the same growth methods as the low-temperature growth of silicon. The gas used to introduce the germanium into the layers is germane GeH_4. The influence of germanium on the growth rate is complex, as illustrated in Figure 7.17. At temperatures in the range 577–650°C, a

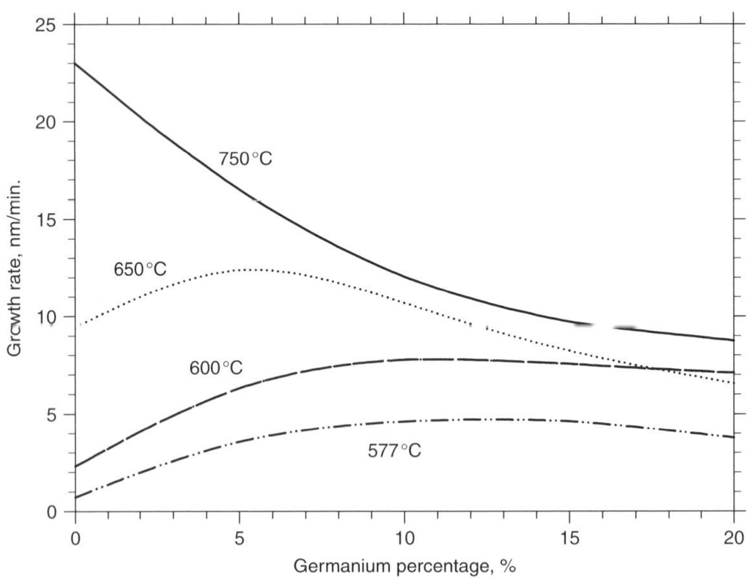

Figure 7.17 Growth rate of $Si_{1-x}Ge_x$ as a function of germanium percentage for temperatures in the range 577–750°C (reproduced with permission from the American Institute of Physics [25])

peak in the growth rate is seen. At low germanium contents, the growth rate increases with germanium content, whereas at high germanium content, the growth rate decreases with germanium content. In the low-temperature regime it has been proposed that hydrogen desorption from the surface is the rate-limiting step. In $Si_{1-x}Ge_x$ this occurs more easily at germanium sites than at silicon sites and hence the growth rate increases with germanium content [26]. As the germanium content increases, the surface contains more and more germanium and less and less hydrogen. The rate limiting step then becomes the adsorption of germane or silane. Robbins [27] proposed that the sticking coefficient for germane or silane was lower at germanium sites. This would slow the adsorption rate as the germanium content increased and hence slow the growth rate.

7.7 SELECTIVE EPITAXY

Selective epitaxy is the growth of a single-crystal layer in a window, with complete suppression of growth elsewhere, and it can be achieved in a number of different ways. The most common method of achieving both selective Si and $Si_{1-x}Ge_x$ epitaxy is by introducing chlorine or HCl into the growth chamber. This can either be done by adding chlorine or HCl as a separate gas or by using a growth gas that contains chlorine, for example dichlorosilane, SiH_2Cl_2. With chlorine chemistry, selective growth of silicon and $Si_{1-x}Ge_x$ can be achieved to both silicon dioxide and silicon nitride.

Chlorine is reported to have two effects that lead to selective growth. First it increases the surface mobility of silicon and germanium atoms, so that atoms deposited on the oxide or nitride layer are able to diffuse across the surface to the window where the growth is occurring. Second it acts as an etch [22] and hence can remove silicon or germanium atoms deposited on the oxide or nitride. The strength of the etching action increases with chlorine content, and if the chlorine content is too high etching of the substrate will occur instead of epitaxial growth.

A typical growth process for selective silicon epitaxy would use silane and a few percent of HCl [22]. The growth rate for this process is shown in Figure 7.16, and compared with the growth rate for dichlorosilane and silane epitaxy. It can be seen that the activation energy for the silane plus HCl process is very similar to that for the dichlorosilane process, indicating that the growth mechanisms are similar. One disadvantage of chlorine-based growth processes over the silane process is a lower growth rate at low temperatures, as can clearly be seen in Figure 7.16.

It is also possible to grow silicon selectively using dichlorosilane and HCl [28].

$Si_{1-x}Ge_x$ can be selectively grown using very similar growth processes to those used for selective silicon epitaxy. In fact the growth of selective $Si_{1-x}Ge_x$ is generally easier to achieve than the growth of selective silicon, as illustrated in Figure 7.18 [29] for $Si_{1-x}Ge_x$ growth using germane and dichlorosilane. It can be seen that the growth moves from non-selective to selective as the proportion of germane in the gas flow increases.

Arrhenius plots for $Si_{1-x}Ge_x$ growth using germane and dichlorosilane are shown in Figure 7.19 for $Si_{1-x}Ge_x$ layers grown using germane and dichlorosilane and for two different HCl flows. It can be seen that the growth rate decreases and the activation energy increases with increasing HCl flow. The explanation proposed for this behaviour is that the limiting growth mechanism changes from hydrogen desorption from the growing surface to chlorine or HCl desorption from the surface [30]. This decrease in growth rate at high HCl flows is a disadvantage for the production of $Si_{1-x}Ge_x$ HBTs because it leads to increased growth times. High HCl flows can also cause surface roughening when the $Si_{1-x}Ge_x$

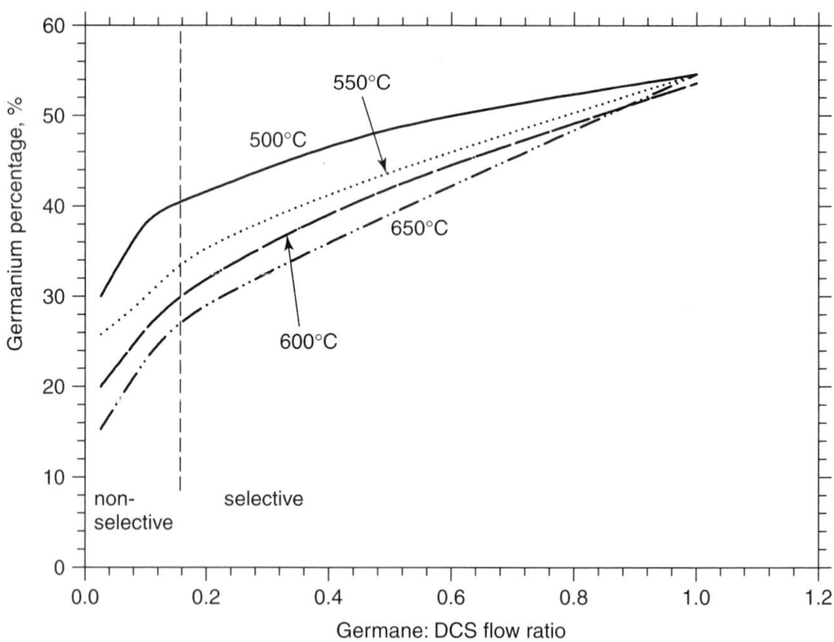

Figure 7.18 Germanium percentage as a function of germane:DCS flow ratio for temperatures in the range 500–650°C showing the move from non-selective to selective growth as the proportion of germane in the gas flow increases (after [29])

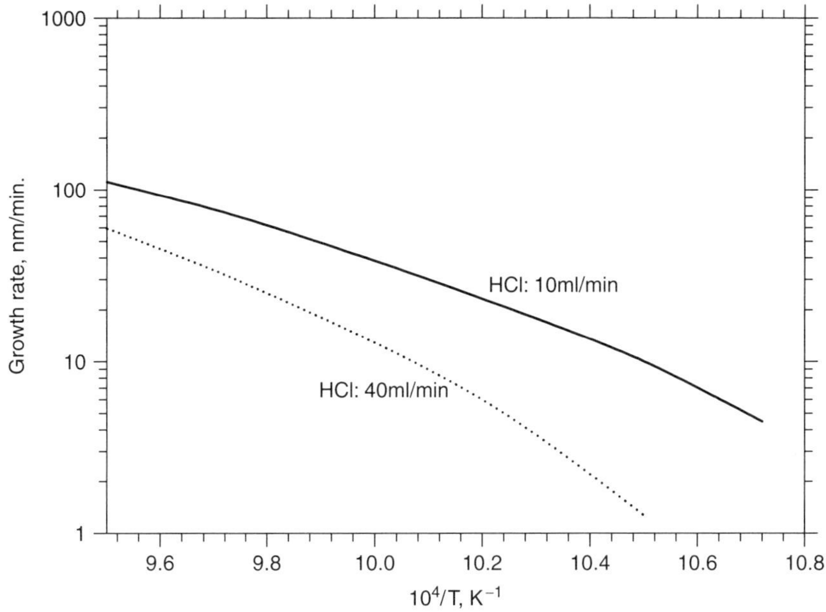

Figure 7.19 Arrhenius plots for $Si_{1-x}Ge_x$ growth at two different HCl flow rates. The dichlorosilane and germane flow rates were fixed at 100 and 8 ml/min respectively (reprinted with permission from [30])

layer is heavily boron doped [30]. These considerations demonstrate that the HCl flow should be chosen to be to the smallest value that is consistent with good selective epitaxy.

Silane can be used for selective silicon epitaxy if the growth is performed at a high temperature. This approach relies on the fact that nucleation of growth on oxide is more difficult than that on silicon. This incubation time for growth on an oxide layer is relatively long at high temperatures but much shorter for growth at low temperatures. Selective silicon layers 1 μm thick can be grown using silane at a temperature of 960°C [31], but the achievable layer thickness decreases with decreasing temperature. At 800°C the maximum selective silicon layer thickness is around 130 nm, at 700°C it is around 60 nm and at 620°C it is around 40 nm. Selective growth to silicon dioxide can be achieved using silane only, but not to silicon nitride.

7.7.1 Faceting and Loading Effects

Selective epitaxy has two process control problems that make it difficult to implement in production technologies, namely faceting and loading

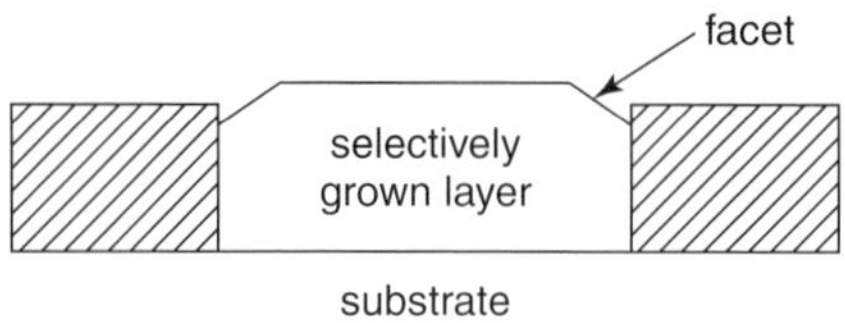

Figure 7.20 Schematic view of a selectively grown layer, showing the formation of facets at the periphery of the growth window

effects. Faceting gives a non-planar surface after growth, as illustrated in Figure 7.20. On a (100) substrate, with windows aligned along the usual (110) direction, the facets form an angle of about 23° with the (100) surface and hence have been assigned as (311) facets. Facet formation is less severe when windows are aligned along the (100) direction (45° to the flat on a (100) wafer). The explanation for facet formation is different growth rates on different wafer orientations. Facet formation can be minimized by optimizing the temperature, pressure and HCl content of the selective growth in the following way:

- reducing the growth rate by growing at a lower temperature;
- reducing the growth rate by growing at a lower pressure;
- increasing the surface mobility by adding chlorine to the source gas;
- increasing the surface mobility by adding HCl.

Loading effects are commonly seen in selective epitaxy, and comprise different layer thicknesses in different window sizes and a dependence of layer thickness on the fraction of the surface covered by oxide. Loading effects arise because of the way that selective epitaxy is achieved through the use of a high surface mobility for deposited atoms. This inevitably leads to a dependence of the layer thickness on the amount of oxide or nitride surrounding the growth window. Loading effects can either lead to an enhancement or retardation of the growth rate in small windows. In silicon selective epitaxy, the growth rate tends to be retarded in small windows whereas in selective $Si_{1-x}Ge_x$ epitaxy it tends to be enhanced. HCl tends to worsen loading effects because it enhances the surface mobility of silicon and germanium atoms. Fortunately, loading effects tend to saturate in very small windows, so loading effects can be controlled, provided the growth process is developed on wafers containing a wide range of different window sizes.

REFERENCES

[1] R. Braunstein, A.R. Moore and F. Herman, 'Intrinsic optical absorption in germanium-silicon alloys', *Phys. Rev.* **109**, 695 (1958).

[2] S.S. Iyer, G.L. Patton, J.M.C. Stork, B.S. Meyerson and D.L. Harame, 'Heterojunction bipolar transistors using Si-Ge alloys', *IEEE Trans. Electron. Devices*, **36**, 2043 (1989).

[3] C.A. King, J.L. Hoyt and J.F. Gibbons, 'Bandgap and transport properties of SiGe by analysis of nearly ideal Si/SiGe/Si heterojunction bipolar transistors', *IEEE Trans. Electron. Devices*, **36**, 2093 (1989).

[4] J.M. Matthews and A.E. Blakeslee, 'Defects in epitaxial multilayers I Misfit dislocations in layers', *Jnl Crystal Growth*, **27**, 118 (1974).

[5] J.M. Matthews and A.E. Blakeslee, 'Defects in epitaxial multilayers III Preparation of almost perfect multilayers', *Jnl Crystal Growth*, **32**, 265 (1975).

[6] R. People and J.C. Bean, 'Calculation of critical layer thickness versus lattice mismatch for GeSi/Si strained layer heterostructures', *Appl. Phys. Lett*, **47**, 322 (1985).

[7] S.C. Jain, T.J. Gosling, J.R. Willis, R. Bullough and P. Balk, 'A theoretical comparison of the stability characteristics of capped and uncapped GeSi strained epilayers', *Solid State Electronics*, **35**, 1073 (1992).

[8] S. Margalit, A. Bar-lev, A.B. Kuper, H. Aharoni and A. Neugroschel, 'Oxidation of SiGe alloys', *Jnl Crystal Growth*, **17**, 288 (1972).

[9] O.W. Holland, C.W. White and D. Fathy, 'Novel oxidation processes in Ge implanted Si and its effect on oxidation kinetics', *Appl. Phys. Lett.*, **51**, 520 (1987).

[10] R. People, 'Indirect bandgap of coherently strained SiGe bulk alloys on <011> silicon substrates', *Phys. Rev. B*, **32**, 1405 (1985).

[11] J. Poortmans, S.C. Jain, D.H.J. Totterdell, M. Caymax, J.F. Nijs, R.P. Mertens and R. Van Overstraeten, 'Theoretical calculation and experimental evidence of the real and apparent bandgap narrowing due to heavy doping in *p*-type silicon and strained $Si_{1-x}Ge_x$ layers', *Solid State Electronics*, **36**, 1763 (1993).

[12] T. Manku and A. Nathan, 'Effective mass for strained p-type $Si_{1-x}Ge_x$', *Jnl Appl. Phys*, **69**, 8414 (1991).

[13] T. Manku and A. Nathan, 'Energy band structure for strained *p*-type $Si_{1-x}Ge_x$', *Phys. Rev. B*, **43**, 12634 (1991).

[14] J. Poortmans, 'Low temperature epitaxial growth of silicon and strained $Si_{1-x}Ge_x$ layers and their application in bipolar transistors', PhD thesis, University of Leuven (1993).

[15] J.M. McGregor, T. Manku, A. Nathan, 'Measured in-plane hole drift mobility and Hall mobility in heavily doped, strained *p*-type $Si_{1-x}Ge_x$', presented at Electronic Materials Conference, Boston (1992).

[16] M. Meuris, S. Verhaverbeke, P.W. Mertens, M.M. Heyns, L. Hellemans, Y. Bruynseraede and A. Philipessian, 'The relationship of the silicon surface roughness and gate oxide integrity in NH_4OH/H_2O_2 mixtures', *Japanese Jnl Appl. Phys.*, **31**, L1514 (1992).
[17] A. Ishizaki and YShiraki, 'Low temperature surface cleaning of silicon and its application to silicon MBE', *Jnl Electrochem. Soc.*, **129**, 666 (1986).
[18] B.S. Meyerson, F.J. Himpsel and K.J. Uram, 'Bistable conditions for low temperature silicon epitaxy', *Appl. Phys. Lett.*, 57, 1034 (1990).
[19] G.S. Higashi, Y.T. Chabal, G.W. Trucks and K. Raghavachari, 'Ideal hydrogen termination of the silicon surface', *Appl. Phys. Lett.*, **56**, 656 (1990).
[20] F.W. Smith and G. Ghidini, 'Reaction of oxygen with Si (111) and (100): critical conditions for growth of SiO_2', *Jnl Electrochem. Soc.*, **129**, 1300 (1982).
[21] G. Ghidini and F.W. Smith, 'Interaction of H_2O with Si (111) and (100): critical conditions for growth of SiO_2', *Jnl Electrochem. Soc.*, **131**, 2924 (1984).
[22] J.L. Regolini, D. Bensahel, E. Scheid and J. Mercier, 'Selective epitaxial silicon growth in the 650–1100°C range in a reduced pressure chemical vapour deposition reactor using dichlorosilane', *Appl. Phys. Lett.*, **54**, 658 (1989).
[23] G.R. Srinivasan and B.S. Meyerson, 'Current status of reduced temperature silicon epitaxy by chemical vapour deposition', *Jnl Electrochem. Soc.*, **134**, 1518 (1987).
[24] M. Racanelli, D.W. Greve, M.K. Hatalis and L.J. van Yzendoorn, 'Alternative surface cleaning approaches for ultra high vacuum chemical vapour deposition of Si and GeSi', *Jnl Electrochem. Soc.*, **138**, 3783 (1991).
[25] M. Racanelli and D.W. Greve, 'Temperature dependence of the growth of SiGe by ultra-high vacuum chemical vapour deposition', *Appl. Phys. Lett.*, **56**, 2524 (1990).
[26] B.S. Meyerson, K.J. Uram and F.K. LeGoues, 'Co-operative growth phenomena in silicon-germanium low temperature epitaxy', *Appl. Phys. Lett.*, **53**, 2555 (1988).
[27] D.J. Robbins, J.L. Glasper, A.G. Cullis and W.Y. Leong, 'A model for herogeneous growth of SiGe films from hydrides', *Jnl Appl. Phys.* **69**, 3729 (1991).
[28] A. Ishitani, H. Kitajima, N. Endo and N. Kasai, 'Silicon selective epitaxial growth and electrical properties of epi/sidewall interfaces,' *Japanese Jnl Appl. Phys.*, **28**, 841 (1989).
[29] Y. Zhong, M.C. Ozturk, D.T. Grider, J.J. Wortman and M.A. Littlejohn, 'Selective low pressure chemical vapour deposition of $Si_{1-x}Ge_x$ alloys in a rapid thermal processor using dichlorosilane and germane', *Appl. Phys. Lett.*, **57**, 2092 (1990).
[30] Y. Kiyota, T. Udo, T. Hashimoto, A. Kodama, H. Shimamoto, R. Hayami, E. Ohue and K. Washio, 'HCl free selective epitaxial SiGe growth by

LPCVD for high frequency HBTs', *IEEE Trans. Electron. Devices*, **49**, 739 (2002).

[31] J.M. Bonar, 'Process development and characterisation of silicon and silicon-germanium grown in a novel single-wafer LPCVD system', PhD thesis, University of Southampton (1996).

8

Silicon-Germanium Heterojunction Bipolar Transistors

8.1 INTRODUCTION

The design of bipolar transistors requires trade-offs between a number of competing mechanisms. To achieve a fast base transit time, and hence a high value of cut-off frequency, the basewidth needs to be very small, as shown in equation (5.4). The mechanism that limits the extent that the basewidth can be reduced is punch-through of the base, which occurs when the emitter/base depletion region intersects the collector/base depletion region in the base, as discussed in Section 4.6.1. Thinner depletion regions can be achieved by increasing the base doping concentration, and hence one strategy for improving the performance of silicon bipolar transistors would be to increase the base doping concentration, so that narrower basewidths could be achieved without encountering punch-through. The problem with this strategy is that increasing the base doping, degrades the gain, as can be seen from equation (2.43). This trade-off between gain and base transit time is the main issue that limits the maximum achievable cut-off frequency of a silicon bipolar transistor. In practice, it is technologically difficult to obtain cut-off frequencies much higher than 50 GHz in silicon bipolar transistors.

As discussed in Chapter 7, SiGe has a lower bandgap than Si, and hence if a bipolar transistor could be created with SiGe in the base and

SiGe Heterojunction Bipolar Transistors Peter Ashburn
 ISBN: 0-470-84838-3

Si in the emitter, the theory in Section 3.3 indicates that much higher values of gain would be achieved. This bandgap engineering introduces a new degree of freedom in the design of bipolar transistors that makes it possible to increase the base doping and reduce the basewidth, while at the same time achieving a reasonable value of gain. In this way, much higher values of cut-off frequency can be achieved with silicon germanium heterojunction bipolar transistors (SiGe HBTs) than Si bipolar junction transistors (Si BJTs). SiGe HBTs have been produced with a cut-off frequency f_T as high as 350 GHz and a maximum oscillation frequency f_{max} as high as 260 GHz [1].

In this chapter, the bandgap engineering of SiGe HBTs will be explained and design equations for SiGe HBTs derived. Fortunately, the theory of silicon bipolar transistors in Chapters 2–5 is directly applicable, and hence only minor modifications to this theory are needed to describe the behaviour of SiGe HBTs. The use of carbon to suppress boron diffusion in SiGe is discussed in the second part of this chapter.

8.2 BANDGAP ENGINEERING

A SiGe HBT is produced by sandwiching a SiGe base between a Si collector and a Si emitter. To understand the physical behaviour of SiGe HBTs, the band diagrams of a SiGe HBT and a Si BJT are compared in Figure 8.1. The band diagram of the SiGe HBT is indicated by the solid line and that for the Si BJT by the dashed line. In the valence band, the bandgap difference is seen as discontinuities at the emitter/base and collector/base heterojunctions, while in the conduction band it is seen as spikes. As discussed in Chapter 7, the majority of the bandgap difference between SiGe and Si occurs in the valence band, so the valence band discontinuity is much bigger than the conduction band spike. In Figure 8.1, the size of the conduction band spike has been exaggerated for clarity, but for most practical purposes the conduction band spike is so small that it has little effect on the electrical behaviour of SiGe HBTs.

A comparison of the band diagrams in Figure 8.1 shows that the barrier height to electron flow from emitter to base E_b (conduction band barrier) is much smaller in the SiGe HBT than the Si BJT. This means that the collector current at a given base/emitter voltage will be bigger in a SiGe HBT than in a Si BJT. The barrier height to hole flow from the base to the emitter (valence band barrier) is approximately the same in the SiGe HBT and the Si BJT, which means that the base currents of the two types of device will be approximately the same. The Gummel plots

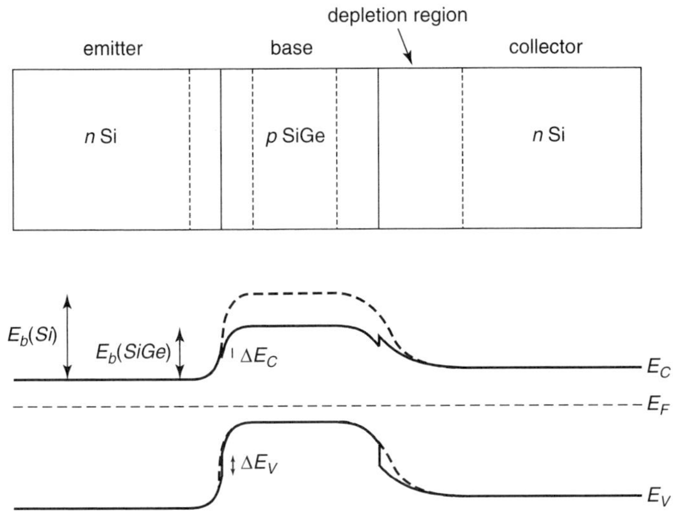

Figure 8.1 Comparison of the band diagrams of a SiGe HBT (solid line) and a Si bipolar transistor (dashed line)

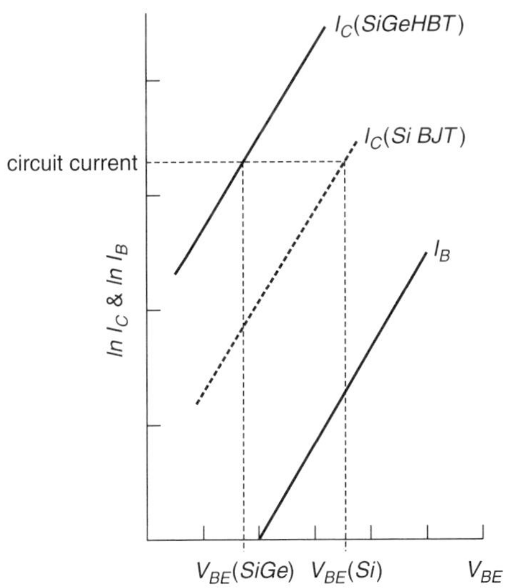

Figure 8.2 Comparison of Gummel plots of a SiGe HBT and a Si bipolar transistor showing the lower V_{BE} for the SiGe HBT

of a comparable SiGe HBT and Si BJT are shown in Figure 8.2. It can be seen that the gain of the HBT is much higher than that of the BJT and that this increased gain is due to an increased collector current. The increased collector current of a SiGe HBT can be thought of in another way. When HBTs are used in circuits, the circuits are usually designed to operate at a given current. If a SiGe HBT and Si BJT are compared at a given current, the HBT has a lower V_{BE}, as illustrated in Figure 8.2. This lower V_{BE} in SiGe HBT circuits is very valuable, since it leads to lower power consumption.

8.3 COLLECTOR CURRENT, BASE CURRENT AND GAIN ENHANCEMENT

Since the conduction band spike in a SiGe HBT is very small, the theory in Section 2.5 can be directly applied to calculate the collector current of a SiGe HBT. By including the effects of doping-induced bandgap narrowing, an equation analogous to equation (3.19) can be derived for the collector current of a SiGe HBT:

$$I_C = \frac{qA(D_{nb})_{SiGe}(n_i^2)_{SiGe}}{W_B N_{aeff}} \exp\frac{qV_{BE}}{kT} \tag{8.1}$$

where it has been assumed that N_{aeff} is the same in SiGe and Si, as discussed in Section 7.3.3. The intrinsic carrier concentration for SiGe can be written as:

$$(n_i^2)_{SiGe} = n_{io}^2 \exp\frac{\Delta E_G}{kT}\left[\frac{(N_C N_V)_{SiGe}}{(N_C N_V)_{Si}}\right] \tag{8.2}$$

In this equation, the germanium-induced bandgap narrowing has been treated in the same way as the doping-induced bandgap narrowing considered in Section 3.3. The term in square brackets corrects for the difference in density of states in SiGe and Si, as discussed in Section 7.3.2. Substituting equation (8.2) into equation (8.1) gives:

$$I_C = \frac{qAD_{nb}n_{io}^2}{W_B N_{ab}} \exp\frac{\Delta E_{gb}}{kT} \exp\frac{qV_{BE}}{kT}\left[\frac{(N_C N_V D_{nb})_{SiGe}}{(N_C N_V D_{nb})_{Si}} \exp\frac{\Delta E_G}{kT}\right] \tag{8.3}$$

where the terms outside the square brackets represent the collector current for a Si BJT, and the terms in square brackets define the correction factor required for a SiGe HBT.

As discussed in Section 8.2, the base current of a SiGe HBT is the same as that for a Si bipolar transistor, and hence the gain enhancement obtainable from a SiGe HBT can be obtained by taking the ratio of collector currents:

$$\frac{\beta_{SiGe}}{\beta_{Si}} = \frac{(N_C N_V D_{nb})_{SiGe}}{(N_C N_V D_{nb})_{Si}} \exp\frac{\Delta E_G}{kT} \tag{8.4}$$

8.4 CUT-OFF FREQUENCY

The cut-off frequency of a SiGe HBT is given by the same equation as for a Si BJT, namely equation (5.20). However, there are some small modifications to the equations for the components of the forward transit time τ_F.

The base transit time τ_B for a SiGe HBT can be calculated using the method in Section 5.2.2. For a uniform base doping profile and a uniform germanium concentration across the base, this method gives an equation analogous to equation (5.4):

$$\tau_B = \frac{W_B^2}{2(D_{nb})_{SiGe}} = (\tau_B)_{Si}\left[\frac{(D_{nb})_{Si}}{(D_{nb})_{SiGe}}\right] \tag{8.5}$$

where the term in square brackets defines the correction that needs to be applied to account for the difference in mobility between SiGe and Si. In practice, this term has a value between 0.83 and 0.71, as discussed in Section 7.3.4.

The emitter delay of a SiGe HBT can be calculated using the method in Section 5.2.3. The charge in the emitter Q_e can be calculated from Figure 5.1 as:

$$Q_e \approx qA\frac{1}{2} W_E p_{eo} \exp\frac{qV_{BE}}{kT} \tag{8.6}$$

The emitter delay τ_E is then defined as:

$$\tau_E = \frac{Q_e}{I_C} = qA\frac{1}{2} W_E p_{eo} \exp\frac{qV_{BE}}{kT}\frac{1}{I_C} \tag{8.7}$$

Since the collector current I_C for a SiGe HBT is much larger than that of a Si BJT, equation (8.7) shows that the emitter delay of a SiGe HBT should be much smaller than that of an equivalent Si BJT. Equation (8.7) can be simplified by substituting equation (8.3) into (8.7):

$$\tau_E = \frac{W_E}{2N_{deff}} \frac{W_B N_{aeff}}{D_{nb}\left[\frac{(N_C N_V D_{nb})_{SiGe}}{(N_C N_V D_{nb})_{Si}} \exp\frac{\Delta E_G}{kT}\right]} \tag{8.8}$$

In this equation, the terms outside the square bracket represent the emitter delay of a Si BJT as given in equation (5.8), and the terms inside the square bracket represent the correction factor required for a SiGe HBT.

8.5 DEVICE DESIGN TRADE-OFFS IN A SiGe HBT

The SiGe base gives new degrees of freedom for the design of SiGe HBTs and allows much higher values of f_T to be achieved than in conventional silicon BJTs. A very high gain is not very useful for most circuit applications, so the approach taken is to trade-off the increased gain of a SiGe HBT for increased base doping. This allows the basewidth to be dramatically reduced without encountering problems of punch-through. To maximize the value of f_T, the boron profile in the SiGe base should be made as thin as possible. To maximize the value of f_{max}, the base resistance and collector/base capacitance also have to be minimized, as can be seen from equation (5.23). The extrinsic components of base resistance and collector/base capacitance can be minimized by using self-aligned fabrication techniques, as will be discussed in Chapters 9 and 10. However, there remains a trade-off between basewidth and intrinsic base resistance. To minimize the base resistance, and hence maximize the value of f_{max}, the doping in the base needs to be as high as possible. To simultaneously maximize the values of f_T and f_{max}, it is clear that the boron profile in the base should be as thin and highly doped as possible. The overall optimization of SiGe HBT technology performance will be considered in more detail in Chapter 12.

When combining a highly doped base with a highly doped emitter, it is necessary to consider emitter/base tunnelling leakage [2], which occurs when the doping concentrations on both sides of a *pn* junction are very high. In this situation the depletion region becomes sufficiently narrow for tunnelling to occur, which results in excess leakage current in reverse bias and non-ideal base characteristics in forward bias. Research has shown that tunnelling leakage occurs when the doping concentration on the low doped side of the junction is greater than about $5 \times 10^{18}\,cm^{-3}$ [2]. One method of producing an HBT with a very heavily doped base is therefore to reduce the doping in the emitter to a level at or below $5 \times 10^{18}\,cm^{-3}$ [3], as illustrated in Figure 8.3(a). This low doped emitter allows a very heavily doped base to be produced without encountering any problems with emitter/base tunnelling leakage. The low doped emitter must be relatively thin to avoid unwanted

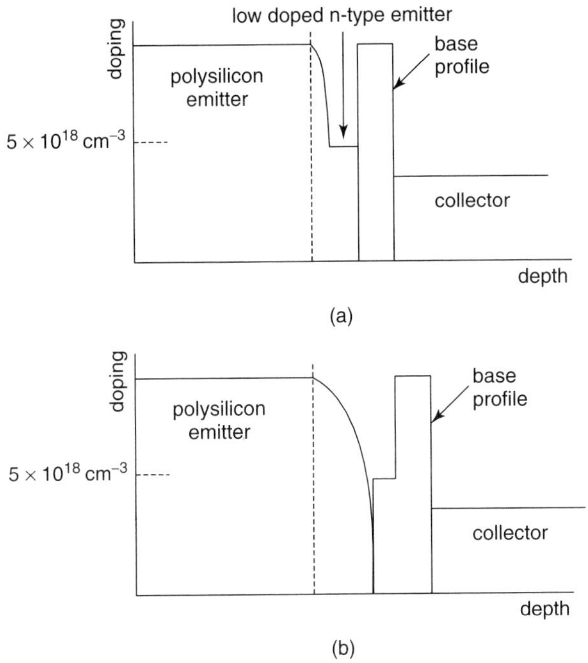

Figure 8.3 Doping profile options for creating a heavily doped base without the problem of emitter/base tunnelling leakage; (a) a low doped emitter; (b) a tailored base profile

stored charge and an increase in the emitter delay, as predicted by equation (8.8). A second approach is to tailor the base profile so that the base doping adjacent to the emitter/base depletion region is less than $5 \times 10^{18}\ \text{cm}^{-3}$, while that deeper in the base is much higher, as illustrated in Figure 8.3(b). The aim here is to give a wide enough emitter/base depletion region to avoid tunnelling, while at the same time minimizing the overall basewidth.

8.6 GRADED GERMANIUM PROFILES

An additional bandgap engineering concept can be applied to further reduce the base transit time and increase the f_T. If the Ge profile is graded across the base, as illustrated in Figure 8.4, the bandgap at the collector is lower than that at the emitter. This gives a gradient on the conduction band, which acts as a built-in electric field, accelerating electrons as they move from the emitter to the collector.

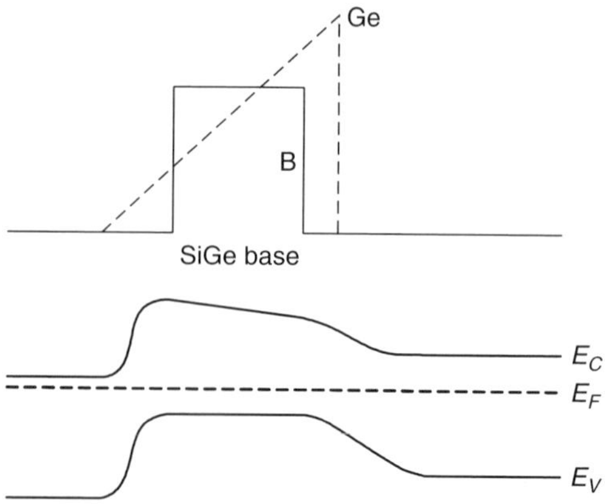

Figure 8.4 Profiles and band diagram of a SiGe HBT with a graded germanium profile

8.6.1 Design Equations for a Graded Germanium Profile

Assuming uniform doping profiles and a linearly graded germanium profile across the base, the equation for the collector current of a graded-base SiGe HBT can be derived as [4]:

$$I_C = \frac{qA\overline{D_{nb}}n_{io}^2}{W_B N_{AB}} \exp\frac{qV_{BE}}{kT} \exp\frac{\Delta E_{gb}}{kT}\left(\frac{(N_C N_V D_{nb})_{SiGe}}{(N_C N_V D_{nb})_{Si}}\right) \times \frac{\Delta E_{G(grade)}}{kT} \frac{\exp(\Delta E_{G(0)}/kT)}{1-\exp-(\Delta E_{G(grade)})/kT} \tag{8.9}$$

where $\Delta E_{G(0)}$ is the germanium-induced bandgap narrowing at the emitter end of the base, $\Delta E_{G(WB)}$ is the germanium-induced bandgap narrowing at the collector end of the base, and $\Delta E_{G(grade)} = \Delta E_{G(WB)} - \Delta E_{G(0)}$ is the grading of the Ge across the base. $\overline{D_{nb}}$ is the average diffusivity of electrons in the graded SiGe base. The gain enhancement in a SiGe HBT with a graded base is then given by:

$$\frac{\beta_{SiGe}}{\beta_{Si}} = \frac{\dfrac{(N_C N_V \overline{D_{nb}})_{SiGe}}{(N_C N_V D_{nb})_{Si}} \dfrac{\Delta E_{G(grade)}}{kT} \exp\dfrac{\Delta E_{G(0)}}{kT}}{1-\exp\dfrac{-\Delta E_{G(grade)}}{kT}} \tag{8.10}$$

This equation indicates that the gain enhancement varies exponentially with the germanium concentration at the emitter end of the base $\Delta E_{G(0)}$, whereas it varies linearly with the grading $\Delta E_{G(grade)}$. This means that if the germanium content is graded from 0% at the emitter end of the base, a relatively small gain enhancement will be obtained. If a large gain enhancement is required, a trapezoidal germanium profile is better than a triangular profile, as shown in Figure 8.5.

The base transit time for a graded-base SiGe HBT can be derived as [4]:

$$\tau_B = \frac{W_B^2}{\overline{D_{nb}}} \frac{kT}{\Delta E_{G(grade)}} \times \left[1 - \frac{kT}{\Delta E_{G(grade)}} \left(1 - \exp - \left(\frac{\Delta E_{G(grade)}}{kT}\right)\right)\right] \tag{8.11}$$

The ratio of base transit times for a graded-base SiGe HBT compared with an equivalent silicon BJT is then given by:

$$\frac{\tau_{BSiGe}}{\tau_{BSi}} = \frac{2kT}{\Delta E_{G(grade)}} \frac{(D_{nb})_{Si}}{(\overline{D_{nb}})_{SiGe}} \left[1 - \frac{kT}{\Delta E_{G(grade)}} \left(\exp - \frac{\Delta E_{G(grade)}}{kT}\right)\right] \tag{8.12}$$

This equation shows that, for a finite germanium grading, this ratio is less than unity, and hence that the grading of the germanium across the base decreases the base transit time and increases the f_T. For a germanium grading across the base of 100 meV, equation (8.12) predicts that the base transit time of a graded-base SiGe HBT is approximately half that of a silicon BJT.

The emitter delay will also be influenced by the grading of the germanium across the base. Equation (8.7) shows that the emitter delay is inversely proportional to the collector current. Substituting equation (8.9) in equation (8.7) gives the following equation for the

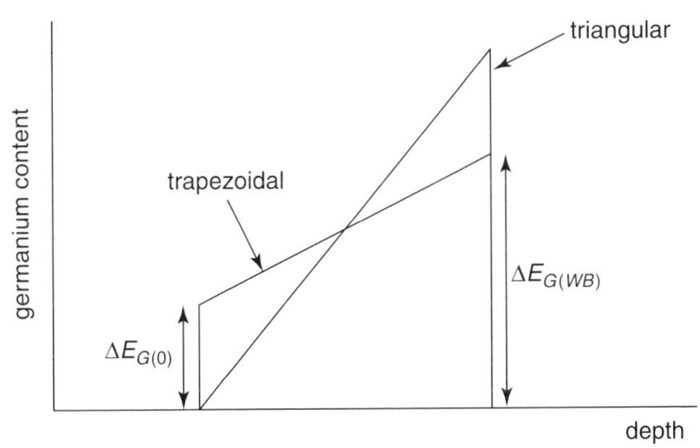

Figure 8.5 Options for germanium profiles in SiGe HBTs with graded germanium

emitter delay of a graded-base SiGe HBT:

$$\tau_E = \frac{W_E}{2N_{deff}} \times \frac{W_B N_{aeff}}{\overline{D_{nb}}\left(\frac{(N_C N_V D_{nb})_{SiGe}}{(N_C N_V D_{nb})_{Si}}\right) \times \frac{\Delta E_{G(grade)}}{kT} \frac{\exp(\Delta E_{G(0)}/kT)}{1-\exp-(\Delta E_{G(grade)})/kT}} \tag{8.13}$$

The ratio of emitter delays for a graded-base SiGe HBT compared with an equivalent silicon BJT is then given by:

$$\frac{\tau_{ESiGe}}{\tau_{ESi}} = \frac{1-\exp\frac{-\Delta E_{G(grade)}}{kT}}{\frac{(N_C N_V \overline{D_{nb}})_{SiGe}}{(N_C N_V D_{nb})_{Si}} \frac{\Delta E_{G(grade)}}{kT} \exp\frac{\Delta E_{G(0)}}{kT}} \tag{8.14}$$

This equation indicates that this ratio decreases exponentially with the germanium concentration at the emitter end of the base $\Delta E_{G(0)}$ but linearly with the grading $\Delta E_{G(grade)}$. This means that if the germanium content is graded from 0% at the emitter end of the base, a relatively small suppression of the emitter delay is obtained. A trapezoidal germanium profile is therefore desirable to maximize the suppression of the emitter delay.

8.7 BORON DIFFUSION IN SiGe HBTs

As discussed in Section 8.5, to achieve high values of f_T and $f_{\max}$ a very thin, heavily doped base is needed. To realize such a base in practice, diffusion of boron must be minimized, both during layer growth and during subsequent high-temperature anneals. In this section, problems associated with boron diffusion in SiGe HBTs are discussed, together with methods for minimizing the diffusion.

8.7.1 Parasitic Energy Barriers

The main criterion that needs to be met for the boron profile in a SiGe HBT is that the boron must be kept within the SiGe layer to achieve full heterojunction action. If the boron penetrates outside the SiGe layer, the metallurgical junction is formed in silicon, rather than SiGe, leading

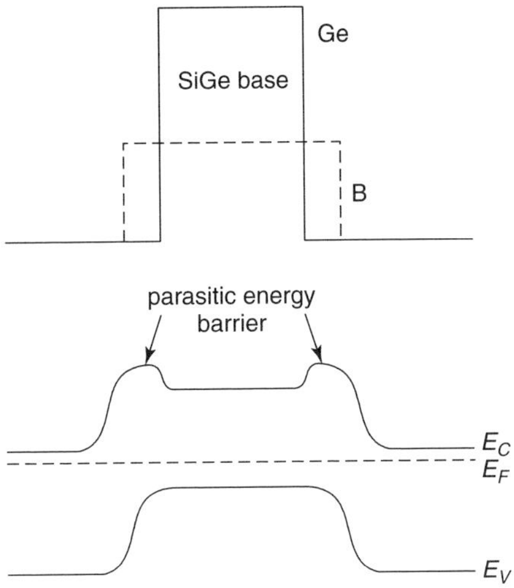

Figure 8.6 Parasitic energy barrier formation due to out-diffusion of boron from a SiGe base

to the formation of parasitic energy barriers [5–9]. This situation is illustrated in the band diagram in Figure 8.6 for an extreme case where a large amount of boron diffusion has occurred into both the emitter and the collector. The metallurgical emitter/base and collector/base junctions are formed in silicon and hence the silicon bandgap is obtained at these junctions. On moving into the SiGe layer, a decrease in bandgap is obtained, which leads to the formation of parasitic energy barriers at both the emitter/base and collector/base junctions. Even very small amounts (a few nanometers) of boron out-diffusion from the SiGe layer dramatically degrade the collector current and hence the gain [7]. The f_T of the HBT is also degraded, since the potential well formed by the parasitic energy barriers traps charge in the base.

A number of useful electrical measurements can be made to detect the presence of parasitic energy barriers in a SiGe HBT. The simplest is to measure the Gummel plots at different values of collector/base reverse bias. If a small amount of boron out-diffusion from the SiGe base is present, increasing the collector/base reverse bias will modulate the parasitic energy barrier at the collector/base junction, reducing the barrier height and giving increased collector current. Measurement of the temperature dependence of the collector current [10] provides an extremely sensitive method of monitoring small amounts of boron out-diffusion from the SiGe. Equation (8.3) shows that the collector current is proportional to the exponential of the bandgap narrowing due to the presence of both heavy boron doping and germanium. An appropriate plot of the temperature dependence of the collector current should therefore give the total bandgap narrowing in the base [10]. If there is some boron out-diffusion from the base, the effective bandgap narrowing due to the germanium will be reduced, thereby reducing the variation of collector current with temperature. This effect is particularly strong when the collector current is measured at low temperatures.

8.7.2 Factors Influencing Boron Diffusion in Si and SiGe

At low doping concentrations, the diffusion coefficient D_i of boron in silicon is constant and is referred to as the intrinsic diffusion coefficient. It can be described by the following simple equation:

$$D_i = D_0 \exp -\frac{E}{kT} \tag{8.15}$$

where the pre-exponential factor D_0 has a value around $0.76\,\text{cm}^2/\text{s}$ and the activation energy E a value of 3.46 eV [11].

In doped silicon, diffusion is more complex, since the diffusion coefficient is not constant, but increases with the doping concentration [11]. The explanation for this effect is related to the detailed atomistic diffusion mechanism, which occurs via point defects, such as vacancies and interstitials [12]. The concentration dependence of the diffusion coefficient arises because the point defect concentrations increase with doping concentration. The precise form of this dependence is determined by the nature of the point defect. In general, point defects can lose an electron and become positively charged (donor type) or gain an electron and become negatively charged (acceptor type). The point defects can also be singly or multiply charged. Taking all these possibilities into account, the diffusion coefficient in doped p-type silicon can be generalized as:

$$D = D_i + D^- \left(\frac{p}{n_i}\right) + D^= \left(\frac{p}{n_i}\right)^2 + D^+ \left(\frac{p}{n_i}\right) + \cdots \tag{8.16}$$

where D_i represents the intrinsic diffusion coefficient for dopant diffusion with a neutral point defect, D^- the intrinsic diffusion coefficient for dopant diffusion with a singly charged acceptor point defect, D^+ the intrinsic diffusion coefficient for dopant diffusion with a singly charged donor point defect, and $D^=$ the intrinsic diffusion coefficient for dopant diffusion with a doubly charged acceptor point defect, etc. The doping dependence of the diffusion coefficient is determined by the terms containing the doping concentration and the intrinsic carrier concentration, which are calculated at the diffusion temperature.

While the full details of atomistic dopant diffusion are not completely understood, it is generally accepted that boron diffusion is primarily determined by an interstitial mechanism and arsenic and antimony diffusion by a vacancy mechanism. Phosphorus diffusion is determined by a mixture of interstitial and vacancy mechanisms, though the interstitial mechanism dominates in most circumstances. Boron diffusion in silicon is thought to involve both neutral and negatively charged dopant-defect pairs [13]. The following dopant-defect interactions are involved:

$$B_s^- + I \Longleftrightarrow (BI)^- \equiv B_i^- \tag{8.17}$$

$$B_s^- + I^+ \Longleftrightarrow (BI) \equiv B_i^0 \tag{8.18}$$

where B_s^- is a substitutional boron atom, I is a neutral silicon self-interstitial and B_i^- is a negatively charged boron-interstitial pair. This first process is termed a kick-out substitutional-interstitial exchange mechanism. Similarly I^+ is a single positively charged silicon self-interstitial and B_i^0 is a neutral boron-interstitial pair.

It would be expected that any process step that influences the interstitial concentration would also have a strong effect on boron diffusion. Ion implantation is an energetic process and hence creates large numbers of point defects as the implanted ion comes to rest in the silicon wafer. Needless to say, the interstitials introduced during ion implantation dramatically increase boron diffusion during high-temperature annealing. For long anneal times, this enhancement of diffusion coefficient is not significant because the interstitials are annealed during the early stages of the anneal. However, for rapid thermal annealing, anneal times of a few seconds are common and in this case a large enhancement of boron diffusion coefficient is obtained. This effect is termed transient enhanced diffusion [14–16] because the enhanced diffusion only occurs in the initial stage of the anneal while the interstitials persist. To model transient enhanced diffusion, the evolution of the spatial point defect concentrations with time must be known, which requires the use of a process simulator. In a SiGe HBT, ion implantation is often used for the selective implanted collector (see Chapter 9) and the extrinsic base [9], and both these process steps can lead to transient enhanced diffusion of boron during later annealing. The polysilicon emitter, though also often doped using ion implantation, does not give rise to transient enhanced diffusion because the implant is made into the polysilicon layer and hence the underlying Si cap and SiGe base remain undamaged.

Diffusion of boron in SiGe has been extensively studied [17–20] and it has been found that boron diffusion is much slower in SiGe than in Si. This is a big advantage for SiGe HBTs, since it is much easier to control boron diffusion in SiGe than it is in Si. The diffusion coefficient of boron has been found to depend almost exponentially on the Ge content of the SiGe layer, and can be described using the simple empirical relation [19,20]:

$$D_{BSiGe}(x) = D_{BSi} \exp -\frac{E_B x}{kT} \tag{8.19}$$

where D_{BSiGe} and D_{BSi} are the boron diffusion coefficients in SiGe and Si respectively, x is the Ge content and the coefficient E_B has a value of 0.7 eV. This equation shows that the boron diffusion coefficient in SiGe decreases with increasing Ge content and that the decrease is a factor of two for an anneal temperature of 1000°C and 11% Ge.

8.7.3 SiGe:C – Reduction of Boron Diffusion by Carbon Doping

Several authors [20–23] have experimentally demonstrated that the introduction of substitutional carbon into Si and SiGe significantly decreases the boron diffusion coefficient. The carbon can be introduced into the base of a SiGe HBT at the same time as the germanium, by using an appropriate gaseous carbon source such as methylsilane. A carbon concentration of around $2 \times 10^{19}\,\text{cm}^{-3}$ has been reported to decrease the boron diffusion coefficient in SiGe by a factor of three over a range of temperatures from 750 to 900°C [24].

The effect of the carbon on boron diffusion is due to the coupled diffusion of carbon and point defects. Carbon diffusion in Si and SiGe occurs by a substitutional-interstitial exchange mechanism in which immobile substitutional carbon atoms C_s are transformed into mobile interstitial carbon atoms C_i via a kick-out mechanism with silicon self-interstitials [20]:

$$C_s + I \Longleftrightarrow C_i \tag{8.20}$$

Furthermore, interstitial carbon can be formed by the Frank-Turnbull dissociative reaction:

$$C_s \Longleftrightarrow C_i + V \tag{8.21}$$

These mechanisms tend to lead to a supersaturation of vacancies and an undersaturation of self-interstitials. The undersaturation of self-interstitials suppresses boron (and phosphorus) diffusion, since the interstitials are not available to contribute to the mechanisms in equations (8.17) and (8.18). Incidentally the supersaturation of vacancies means that carbon enhances the diffusion of arsenic (and antimony) in Si and SiGe [25].

The mechanisms in equations (8.20) and (8.21) are also responsible for the suppression of transient enhanced diffusion that is obtained when carbon is introduced into Si and SiGe [21,25]. A substitutional carbon doping of around 1×10^{20} cm^{-3} is sufficient to suppress transient enhanced diffusion resulting from a phosphorus implant for the selective implanted collector [25].

8.8 STRAIN RELAXATION AND STRAIN COMPENSATED $Si_{1-x-y}Ge_xC_y$

As discussed in Chapter 7, SiGe is a strained, or pseudomorphic, material. It is essential therefore that this strain is maintained during device processing, because if relaxation of the strain occurs, misfit dislocations are generated at the heterojunction interfaces. In a SiGe HBT, these heterojunction interfaces are, of course, located in the emitter/base and collector/base depletion regions. Any relaxation of the strain will therefore lead to generation/recombination in the depletion regions and degradations will be obtained in the base current ideality, the low current gain and the junction leakages, as discussed in Section 4.2.4. When designing the base of a SiGe HBT it is therefore essential that the basewidth and germanium content are chosen so that the basewidth is below the critical thickness of the SiGe layer, as discussed in Chapter 7. If this is done, the SiGe layer will be stable and will not relax during subsequent high-temperature processing. Fortunately this requirement is easy to meet in most SiGe HBTs because a narrow base is needed for other reasons, in particular to achieve a high value of f_T.

An alternative method of managing the strain in a SiGe HBT is to introduce a small percentage of carbon into the SiGe to create a $Si_{1-x-y}Ge_xC_y$ alloy layer. Carbon atoms are smaller than both silicon and germanium atoms and hence substitutional carbon can be used to compensate the strain in a $Si_{1-x}Ge_x$ layer [26]. Full compensation of the strain can be achieved if 1% of substitutional carbon is introduced for every 10% of Ge. This concentration of carbon is approximately ten times higher than is needed for boron diffusion suppression. Strain compensated $Si_{1-x-y}Ge_xC_y$ would have the advantage of removing any possibility of strain relaxation, and hence the wafers could be processed at very high temperatures without fear of strain relaxation. This approach might be attractive for power bipolar transistors where a wider base is needed for reasons of breakdown voltage.

REFERENCES

[1] J.-S.Rieh *et al.*, 'SiGe HBTs with cut-off frequency of 350 GHz', *IEDM Technical Digest*, 771 (2002).

[2] J.A.del Alamo and R.M. Swanson, 'Forward biased tunnelling: a limitation to bipolar device scaling', *IEEE Electron. Device Lett.*, 7, 629 (1986).

[3] A. Schüppen, 'SiGe HBTs for mobile communication', *Solid State Electronics*, **43**, 1373 (1999).

[4] D.L. Harame, J.H. Comfort, J.D. Cressler, E.F. Crabbé, J.Y.-C. Sun, B.S. Meyerson and T. Tice, 'Si/SiGe epitaxial base transistors- part I: materials, physics and circuits', *IEEE Trans. Electron. Devices*, **42**, 455 (1995).

[5] E.J. Prinz, P.M. Garone, P.V. Schwartz, X. Xiao and J.C. Sturm, 'The effect of base dopant out-diffusion and undoped SiGe junction spacer layers in SiGe heterojunction bipolar transistors', *IEEE Electron. Device Lett.*, **12**, 42 (1991).

[6] J.W. Slotboom and G. Streutker, A. Pruijmboom and D.J. Gravesteijn, 'Parasitic energy barriers in SiGe HBTs', *IEEE Electron. Device Lett.*, **12**, 486 (1991).

[7] Z.A. Shafi, P. Ashburn, I.R.C. Post, D.J. Robbins, W.Y. Leong, C.J. Gibbings and S. Nigrin, 'Analysis and modelling of the base currents of Si/SiGe heterojunction bipolar transistors fabricated in high and low oxygen content material', *Jnl Appl. Phys.*, **78**, 2823 (1995).

[8] B. LeTron, Md. R.Hashim, P.Ashburn, M.Mouis, A.Chantre and G.Vincent, 'Determination of bandgap narrowing and parasitic energy barriers in SiGe HBTs integrated in a bipolar technology', *IEEE Trans. Electron. Devices*, **44**, 715 (1997).

[9] Md. R.Hashim, R.F.Lever and P.Ashburn, '2D simulation of the effects of transient enhanced boron out-diffusion from base of SiGe HBT due to an extrinsic base implant, *Solid State Electronics*, **43**, 131 (1999).

[10] P. Ashburn, H. Boussetta, Md. R.Hashim, A.Chantre, M.Mouis, G.J.Parker and G.Vincent, 'Measurement of the bandgap narrowing in the base of Si homojunction and SiGe heterojunction bipolar transistors from the temperature dependence of the collector current', *IEEE Trans. Electron. Devices*, **43**, 774 (1996).

[11] R.B. Fair, 'Concentration profiles of diffused dopants in silicon', in *Impurity Doping Processes in Silicon*, Chapter 7, Ed. F.F.Y. Wang, North Holland, New York (1981).

[12] P.M. Fahey, P.B. Griffin and J.D. Plummer, 'Point defects and dopant diffusion in silicon', *Rev. Modern Phys.*, **61**, 289 (1989).
[13] F.F. Morehead and R.F. Lever, 'Enhanced tail diffusion of phosphorus and boron in silicon: self-interstitial phenomenon', *Appl. Phys. Lett.*, **48**, 151 (1986).
[14] S. Solmi, F. Baruffaldi and R. Canteri, 'Diffusion of boron in silicon during post-implantation annealing', *Jnl Appl. Phys.*, **69**, 2135 (1991).
[15] Y.M. Kim, G.Q. Lo, H. Konoshita and D.L. Kwong, 'Roles of extended defect evolution on the anomalous diffusion of boron in silicon during rapid thermal annealing', *Jnl Electrochem. Soc.*, **138**, 1122 (1991).
[16] R.B. Fair, 'Junction formation in silicon by rapid thermal annealing', in *Rapid Thermal Processing: Science and Technology*, Ed. R.B. Fair, Academic Press (1993).
[17] P. Kuo, J.L. Hoyt, J.F. Gibbons, J.E. Turner, R.D. Jacowitz and T.I. Kamins, 'Comparison of boron diffusion in Si and strained SiGe epitaxial layers', *Appl. Phys. Lett.*, **62**, 612 (1993).
[18] N. Moriya, L.C. Feldman, H.S. Luftman, C.A. King, J. Bevk and B. Freer, 'Boron diffusion in strained SiGe epitaxial layers', *Phys. Rev. Lett.*, **71**, 883 (1993).
[19] N.E.B. Cowern, P.C. Zalm, P. van der Sluis, D.J. Gravensteijn and W.B. de Boer, 'Diffusion in strained SiGe', *Phys. Rev. Lett.*, **72**, 2585 (1994).
[20] H. Rücker and B. Heinemann, 'Tailoring dopant diffusion for advanced SiGe:C heterojunction bipolar transistors', *Solid State Electronics*, **44**, 783 (2000).
[21] L.D. Lanzerotti, J.C. Sturm, E. Stach, R. Hull, T. Buyuklimanli and C. Magee, 'Suppression of boron transient enhanced diffusion in SiGe heterojunction bipolar transistors by carbon incorporation', *Appl. Phys. Lett.*, **23**, 3125 (1997).
[22] H.J. Osten, G. Lippert, D. Knoll, R. Barth, B. Heinemann, H. Rücker and P. Schley, 'The effect of carbon incorporation on SiGe heterojunction bipolar transistor performance and process margin', *IEDM Technical Digest*, 803 (1997).
[23] A. Gruhle, H. Kibbel and U. König, 'The reduction of base dopant out-diffusion in SiGe heterojunction bipolar transistors by carbon doping', *Appl. Phys. Lett.*, **75**, 1311 (1999).
[24] H. Rücker, B. Heinemann, Röpke, R.Kurps, D.Krüger, G.Lippert and H.J.Osten, 'Suppressed diffusion of boron and carbon in carbon-rich silicon', *Appl. Phys. Lett.*, **73**, 1682 (1998).
[25] H. Rücker, B. Heinemann, D. Bolze, D. Knoll, D. Krüger, R. Kurps, H.J. Osten, P. Schley, B. Tillack and P. Zaumseil, 'Dopant diffusion in C-doped Si and SiGe: physical model and experimental verification', *Proc. BCTM* (1999).
[26] A. St. Amour, J.C. Sturm, Y. Lacroix and M.L.W. Thewalt, 'Defect-free band-edge photoluminescence and bandgap measurement of pseudomorphic SiGeC alloy layers on Si (100)', *Appl. Phys. Lett.*, **67**, 3915 (1995).

9

Silicon Bipolar Technology

9.1 INTRODUCTION

The design of bipolar transistors is intimately interwoven with the methods used for their fabrication. Any study of bipolar transistors would therefore be incomplete without consideration of the limitations imposed by the fabrication technology. Furthermore, bipolar transistors are generally incorporated into integrated circuits, and hence the technology must take into account the constraints imposed by the circuit configuration. For high-speed digital circuits and high-frequency analogue circuits, the main requirement is to minimize all parasitic resistances and capacitances. As will be discussed in Chapter 12, the most important parasitics in a bipolar transistor are collector/base capacitance, emitter/base capacitance and base resistance. Self-aligned processing techniques, analogous to those used in CMOS, have been developed to minimize these parasitic resistances and capacitances. These developments have led to the creation of the self-aligned, double-polysilicon bipolar process.

Breakdown voltage is another important parameter that has a strong influence on the design of bipolar transistors. High breakdown voltages imply the use of a thick, lowly doped epitaxial layer in order to give a wide collector/base depletion region. These requirements run counter to those for speed, where shallow junctions are needed to minimize peripheral capacitance and a thin heavily doped epitaxial layer to suppress the Kirk effect, as discussed in Section 5.5. It is therefore difficult to simultaneously achieve high-frequency performance and a high breakdown voltage. The breakdown voltage is primarily determined by the front-end processing, particularly the epitaxy and

SiGe Heterojunction Bipolar Transistors Peter Ashburn
 ISBN: 0-470-84838-3

collector fabrication. The processing of the buried layer and epitaxy will therefore be described in detail, together with the selective implanted collector, which is a method of optimizing both the high-frequency performance and the trade-off with the breakdown voltage.

Bipolar transistors are often combined with MOS transistors in BiCMOS technologies, which has the advantage of allowing MOS and bipolar circuits to be mixed on a single chip. The main advantages of MOS transistors are low power consumption, high packing density and ease of design of ULSI digital systems. However, MOS transistors also have a number of disadvantages, foremost among which are a poor transconductance, a poor drive capability and a limited high frequency performance. Bipolar transistors and SiGe HBTs, on the other hand, have a high transconductance, a large current drive per unit silicon area, a high cut-off frequency and a low $1/f$ noise. Optimization of the overall system design is therefore often easier in a BiCMOS technology, where the separate strengths of MOS and bipolar transistors can be combined to give improved system design. A typical BiCMOS technology will be described and the alternative approaches for overall technology optimization discussed.

The key features of a bipolar process are illustrated in Figure 9.1, which shows a cross-section through a basic integrated circuit transistor. Emitter and base regions are clearly needed for the transistor itself as well as a $p+$ extrinsic base region to decrease the extrinsic base resistance. The emitter normally comprises a polysilicon emitter, so that a shallow emitter/base junction can be realized with a low peripheral emitter/base capacitance. Some form of electrical isolation must be included to prevent unwanted conduction between adjacent transistors.

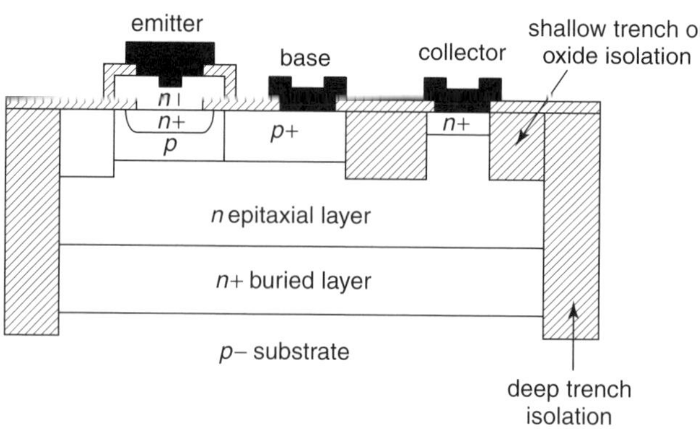

Figure 9.1 Cross-section of a basic, silicon bipolar technology

This is generally achieved by a combination of deep trench isolation and either shallow trench isolation or oxide isolation. A buried layer is desirable to reduce the collector resistance, and this necessitates the use of an epitaxial layer. These are the key elements of any bipolar process, and will be considered in more detail in the following sections.

9.2 BURIED LAYER AND EPITAXY

The relatively low doping concentration in the collector of a bipolar transistor (typically $1 \times 10^{17}\,cm^{-3}$) introduces a large series collector series resistance. This can seriously degrade the electrical performance of the transistor, giving rise to a serious reduction in the current-carrying capability of the transistor and an increase in the saturation voltage. For these reasons, a buried $n+$ layer is incorporated below the active device region, as shown in Figure 9.1. This provides a low-resistance path to the collector contact, thereby short-circuiting the highly resistive epitaxial collector. In some cases, an $n+$ collector sink diffusion, connecting the collector contact and the buried layer, is also included to further reduce the collector resistance.

The process sequence for buried layer and epitaxy is shown in Figure 9.2. The buried layer is fabricated by implanting arsenic or antimony and then heating at a high temperature to diffuse the dopant into the substrate. These dopants are chosen over phosphorus because of their very low diffusion coefficient in silicon. An oxidation is often included as part of the buried layer drive-in to produce a step in the silicon surface, as illustrated in Figure 9.2(b). This step should be large enough to be visible through the epitaxial layer, so that subsequent layers can be aligned to the buried layer (e.g. collector sink and base). The step arises because the oxidation rate of heavily doped $n+$ silicon is much greater than that of lightly doped silicon [1]. The buried layer junction depth is determined by the requirement for a low sheet resistance (typically $10\,\Omega/sq$) to minimize collector resistance, and by the need for a low surface concentration to avoid autodoping [2–4] during epitaxy, as will be discussed shortly. Sheet resistances lower than about $10\,\Omega/sq$ are difficult to achieve because of defect generation during drive-in and epitaxial growth [5,6].

Epitaxy [1] is the term applied to the growth of a single-crystal layer of semiconductor on a single-crystal substrate. The crystalline substrate serves as a seed for the epitaxial growth, and allows the process to take place at a temperature well below the melting point of silicon. A wide range of temperatures can be used for epitaxy (500–1200°C),

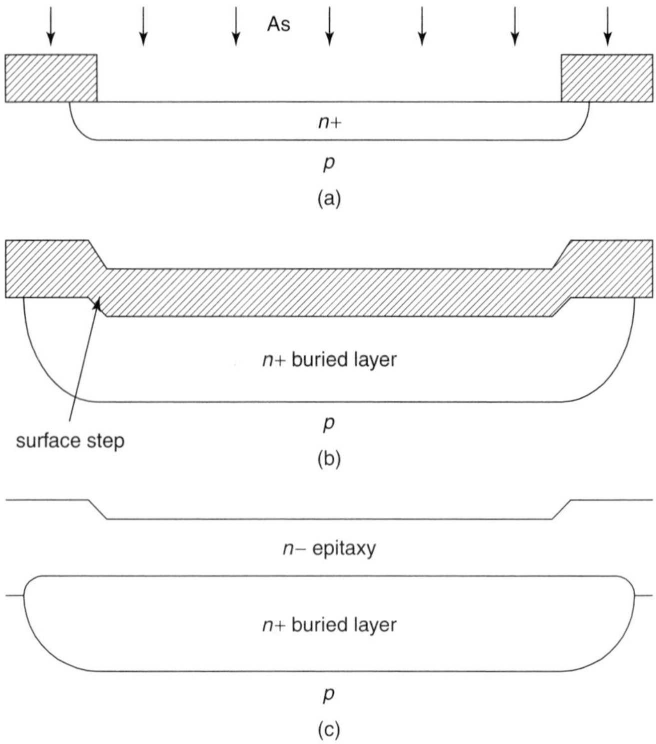

Figure 9.2 Buried layer and epitaxy formation in a bipolar transistor

with lower temperatures being advantageous for thin epitaxial layers and higher temperatures for thick epitaxial layers. The epitaxial process proceeds by the reduction of a gaseous silicon compound such as silane (SiH_4), dichlorosilane (SiH_2Cl_2) or silicon tetrachloride ($SiCl_4$). Dopants can be incorporated into the growing epitaxial layer by mixing the silicon source gas with a gaseous hydride (AsH_3, PH_3 or B_2H_6). As discussed in Chapter 7, the selection of the optimum conditions for epitaxial growth is a very complex decision, based on factors such as reactor configuration, deposition temperature, growth rate, surface cleanliness, etc. The crystalline quality of epitaxial layers is of paramount importance, and defect generation during epitaxy must be avoided. Defects in the substrate, such as dislocations, are able to propagate into the growing epitaxial film, and additional defects such as epitaxial stacking faults [6] can be nucleated at impurities or damage on the substrate surface.

Epitaxial layer thickness can be controlled over a wide range of values. Layers of several hundred microns can be grown for power device applications [7], and layers of less than a micron for high-speed bipolar transistors [8]. In sub-micron epitaxial layers autodoping and out-diffusion of the buried layer are the main factors that constrain the extent that the thickness can be reduced. These mechanisms are illustrated in Figure 9.3, where it can be seen that autodoping is manifested as a tail on the diffusion profile of the buried layer dopant up into the epitaxial layer. Autodoping [2–4] of epitaxial layers occurs through diffusion and evaporation of dopant from the substrate. The dopant is then incorporated into the growing epitaxial layer through the gas phase. This is a particular problem in bipolar processes because of the presence of the heavily doped buried layer. Autodoping can be minimized by ensuring that the buried layer surface concentration is low and by growing the epitaxial layer at a low temperature.

Pattern shift, pattern distortion and pattern washout [9–11] can also often observed in epitaxy processes. Pattern shift is illustrated in Figure 9.4(a), where it can be seen that the depression introduced into the substrate surface during the buried layer drive-in has shifted during epitaxy. Pattern distortion is shown in Figure 9.4(b) and is an effect in which the shape of the depression is distorted during epitaxy. Washout is shown in Figure 9.4(c) and is an effect where the depression is smeared during epitaxy. Experiments have also shown that these effects are significantly reduced if the epitaxy is carried out at a low pressure [12].

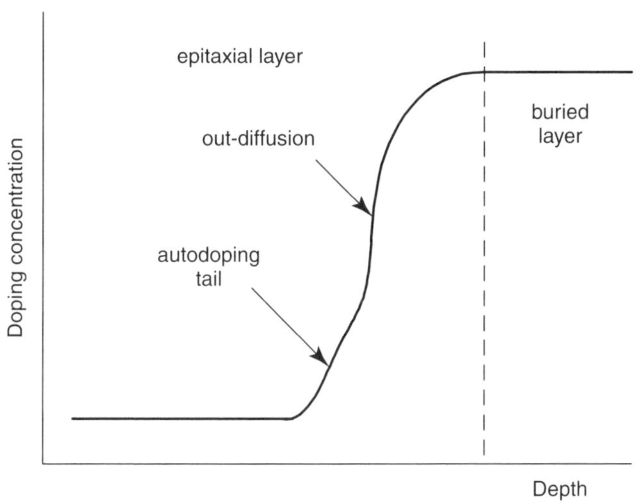

Figure 9.3 Illustration of autodoping and out-diffusion from the buried layer into the epitaxial layer

9.3 ISOLATION

Isolation is needed in integrated circuit processes to ensure that transistors are electrically isolated from other transistors and components in the circuit. The simplest method of isolation for bipolar circuits is junction isolation, as illustrated in Figure 1.1. Electrical isolation between transistors or components in adjacent *n*-epitaxial islands is achieved by reverse biasing the collector/isolation *pn* junction. Since negligible current flows through a reverse-biased *pn* junction, transistors and components in adjacent wells are effectively isolated. The reverse bias is applied by connecting the substrate or isolation regions to the most negative voltage in the circuit. Although this technique is entirely effective, it suffers from the disadvantage of consuming a large amount of silicon area because of the lateral diffusion of the isolation regions. The large parasitic capacitance associated with the collector/isolation *pn* junction also makes it unsuitable for realizing high-speed circuits.

Oxide isolation [13] or shallow trench isolation [14] are commonly used in CMOS technologies to produce the field oxide layer. Both these techniques rely on the use of a recessed silicon dioxide layer to

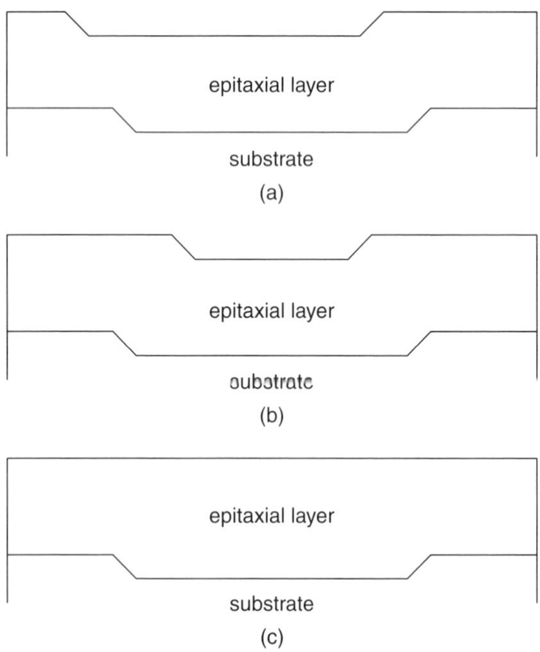

Figure 9.4 Illustration of pattern shift (a), pattern distortion (b) and pattern washout (c)

simultaneously provide a thick silicon dioxide layer and a planar surface. In CMOS processes, a thick field oxide is used in conjunction with a field threshold implant to ensure that parasitic transistors, formed when interconnections pass over the field oxide, do not turn on during circuit operation. Oxide isolation and shallow trench isolation are also used in bipolar technologies, as illustrated in Figures 1.2 and 9.1. However, in bipolar technologies, oxide isolation and shallow trench isolation do not generally provide electrical isolation because the oxide layer is not thick enough to penetrate all the way through the epitaxial layer and buried layer (several microns). In bipolar technologies, oxide isolation and shallow trench isolation are used to improve packing density and reduce parasitic capacitance. The base region can be butted against the recessed oxide layer, as shown in Figure 9.1, which eliminates the sidewall component of the collector/base capacitance and also gives an improvement in packing density. The emitter region can, in principle, also be butted against the recessed oxide to produce a so-called walled emitter, as shown in Figure 9.5. This gives reduced emitter/base capacitance, but unfortunately suffers from yield problems due to the formation of emitter/collector pipes (short circuits between emitter and collector) at defects generated at the vertical oxide/silicon interface [15].

The vast majority of bipolar and BiCMOS technologies use deep trench isolation [16] to provide electrical isolation of bipolar transistors in integrated circuits. Deep trench isolation comprises a deep and narrow trench etched into the silicon, which is filled with an insulator, as illustrated in Figure 9.1. The trench needs to be deep enough to penetrate through both the epitaxial layer and the buried layer. If this is done, no separation is required between adjacent buried layers and hence a big improvement in packing density is obtained. Deep trench isolation can also be used in CMOS technologies [17].

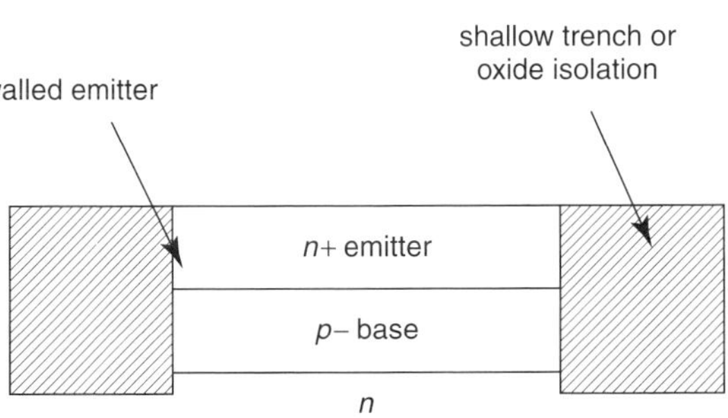

Figure 9.5 Bipolar transistor with a walled emitter

Deep trench fabrication is a three-part process involving trench etching, refilling and planarization, and is illustrated in Figure 9.6. The process sequence starts (Figure 9.6(a)) with the deposition of pad oxide, silicon nitride and a thick masking layer such as photoresist or a deposited silicon dioxide layer. The silicon nitride layer is needed as an etch stop during planarization, and the pad oxide relieves the stress at the nitride/silicon interface. Following photolithography, a deep narrow trench is etched using reactive ion etching (Figure 9.6(b)). The main requirement for the trench is for vertical walls, a criterion that can be easily met when the trench opening is wide but which becomes more difficult to meet as the trench width decreases [18]. A channel-stop implant is generally introduced at this stage of the process to prevent the formation of an *n*-type inversion layer in the underlying lightly doped *p*-type silicon.

Trench refilling can accomplished in a variety of different ways [19,20], but deposition of polysilicon is the most common. The first stage of the refill procedure is generally a thermal oxidation to grow a thin silicon dioxide layer, as shown in Figure 9.6(c). The trench is then refilled by depositing a thick layer of undoped polysilicon. The main criterion that must be met by the refill procedure is the avoidance of defect generation during subsequent heat treatments and oxidations [21,22]. Particular problems arise if the thermal oxide around the inside of the trench is too thick, as well as at the seam in the polysilicon down the centre of the trench. When an oxidation is carried out, an oxide layer can form down the centre of the seam, forcing apart the polysilicon in the trench. This generates a large amount of stress, which is relieved by the formation of dislocations at the corners of the trench.

The final stage of trench isolation is planarization, which is illustrated in Figure 9.6(d). This is achieved by etching back the polysilicon or chemical mechanical polishing to give a planar surface. The silicon nitride layer over the active transistor areas acts as an etch-stop or polish-stop. The deep trench structure is completed by carrying out an oxidation to cap the polysilicon layer at the top of the trench.

Selective epitaxy [23,24] is an alternative approach for isolating bipolar integrated circuit transistors, as illustrated in Figure 9.7. The buried layer is formed by implanting arsenic or antimony through a window in thick silicon dioxide layer, as illustrated in Figure 9.7(a). Following buried layer drive-in, an *n*-type collector is selectively grown using the techniques described in Chapter 7 to produce the structure shown in Figure 9.7(b). The advantage of selective epitaxy is that it eliminates the need for both deep and shallow trench isolation, and hence is a very

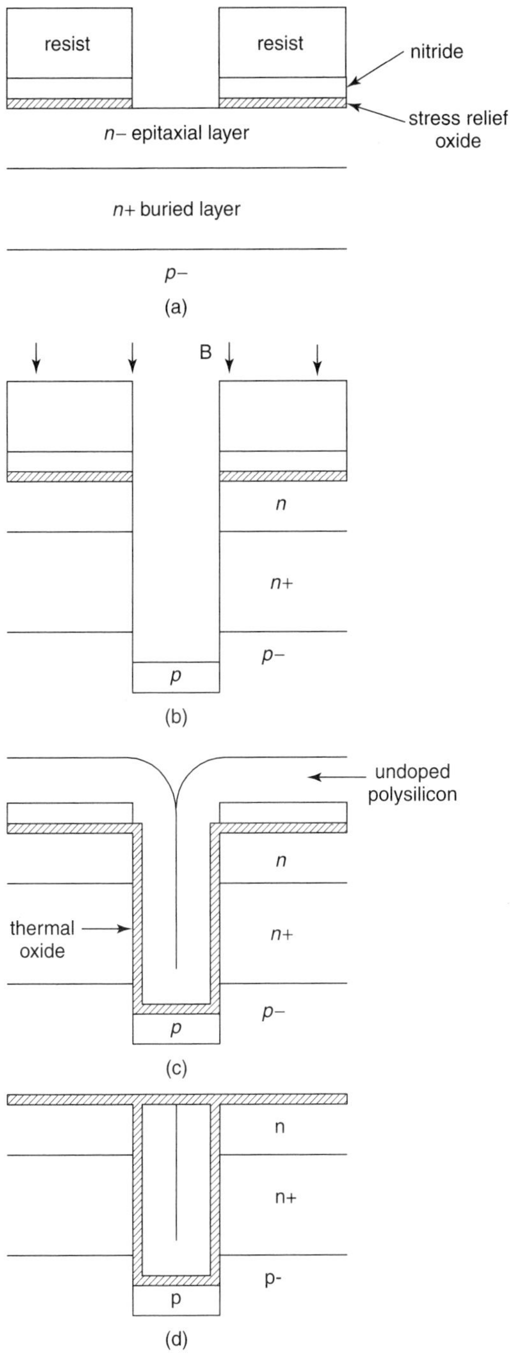

Figure 9.6 Fabrication sequence for deep trench isolation

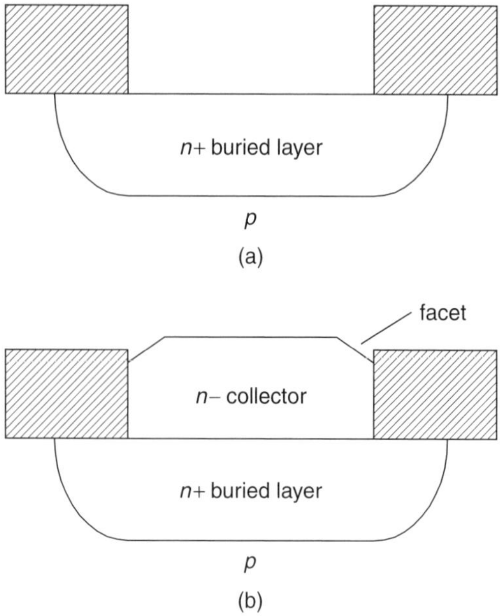

Figure 9.7 Illustration of the use of selective epitaxy for isolation in an integrated circuit bipolar process

simple process. It is particularly attractive for SiGe HBTs, where epitaxy is also needed to produce the SiGe base. The combination of selective silicon epitaxy for the collector and non-selective SiGe epitaxy for the base gives a very simple process for the fabrication of HBTs [25]. The major problems with selective epitaxy are the formation of facets [26] around the periphery of the epitaxial regions and loading effects, which give different thicknesses of silicon for different window sizes [27], as discussed in Section 7.7.

9.4 SELECTIVE IMPLANTED COLLECTOR

To achieve a high value of f_T in a bipolar transistor, a high collector doping concentration is needed to push the Kirk effect to higher currents as discussed in Section 5.5. However, a high collector doping concentration gives a narrow collector/base depletion width, which then gives a high collector/base capacitance. As will be seen in Chapter 12, collector/base capacitance is one of the most important parasitic capacitances in a bipolar transistor and it has a strong effect on the circuit performance. It also degrades the achievable f_{max} of a bipolar transistor, as can be seen from equation (5.23). Some method is therefore needed to manage the trade-off between f_T and collector/base capacitance.

The Selective Implanted Collector (SIC) provides a method of increasing the collector doping in the intrinsic collector, while maintaining a lower collector doping in the extrinsic collector. In this way, it is possible to optimize the trade-off between f_T and collector/base capacitance. The selective implanted collector is realized with a high-energy phosphorus implant through the emitter window, as illustrated in Figure 9.8. This implant increases the doping in the intrinsic collector, while leaving the doping in the extrinsic collector unchanged. It is a self-aligned process because the SIC implant is made through the same window as the emitter.

A typical doping profile for a selective implanted collector is illustrated in Figure 9.9. The energy of the SIC implant is chosen to put the peak of the phosphorus implant just beyond the base and the dose is chosen to control the Kirk effect. Typically the phosphorus concentration at the collector/base junction is around 1×10^{17} cm^{-3}. The selective implanted collector also has the advantage of compensating the tail of the boron profile and hence reducing the basewidth.

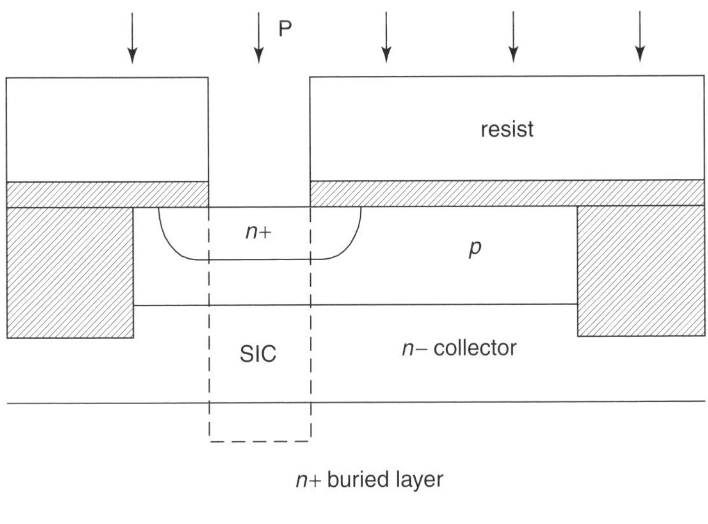

Figure 9.8 Fabrication of a Selective Implanted Collector (SIC) using a phosphorus implant through the emitter window

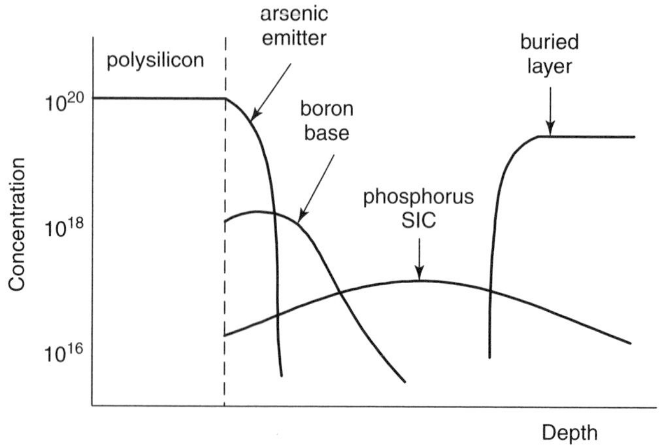

Figure 9.9 Typical doping profiles for a selective implanted collector

9.5 DOUBLE-POLYSILICON, SELF-ALIGNED BIPOLAR PROCESS

The development of the double-polysilicon, self-aligned bipolar process [28–31] led to dramatic improvements in circuit performance. State of the art double-polysilicon, self-aligned bipolar transistors can be produced with values of f_T and f_{max} approaching 70 GHz [8,32,33]. Emitter coupled logic circuits have been produced using these transistors with a gate delay as low as 12 ps and dynamic frequency dividers have been designed that operate at 52 GHz [8]. These impressive results have been made possible by the use of $p+$ polysilicon as a base contact, and by the self-alignment of the emitter to the extrinsic base [28]. The use of $p+$ polysilicon for the base contact gives a reduction in collector/base capacitance, and can be understood by referring to Figure 9.10, which compares a double-polysilicon self-aligned transistor with a conventional transistor. In the conventional transistor the size of the collector/base junction is limited by the requirement to make contact to the emitter and base. The design rules for minimum contact window size, minimum metal-to-metal separation and minimum metal overlap around the contact window therefore determine the size of the collector/base capacitance. In the self-aligned transistor, contact to the base is made via the $p+$ polysilicon layer, so the shallow trench isolation regions can be brought closer together, as shown in Figure 9.10(a). The size of the extrinsic collector/base junction is then limited only by the requirement to provide an overlap between the $p+$ polysilicon and the single-crystal silicon. These changes in the layout of the transistor make possible a

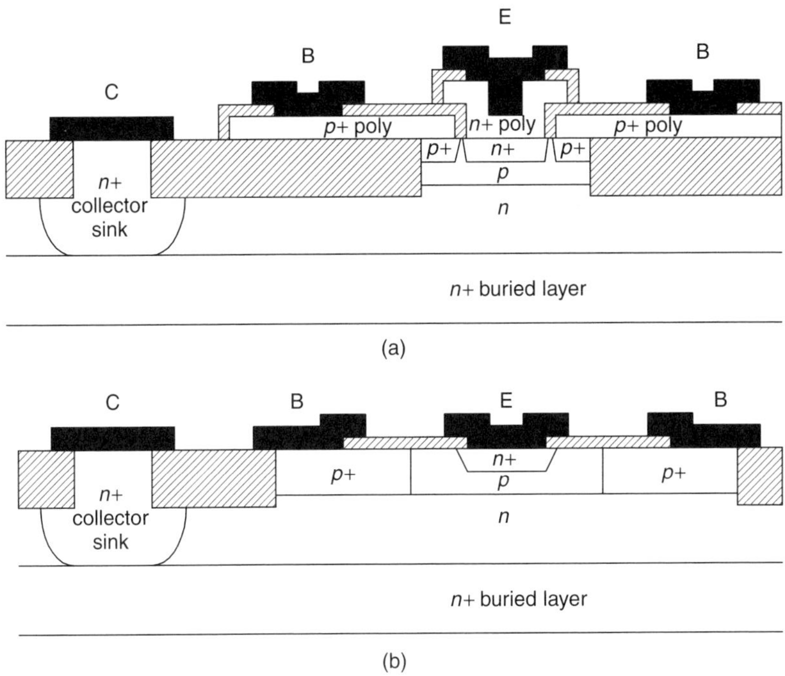

Figure 9.10 Cross-sectional views of a double-polysilicon, self-aligned bipolar transistor (a) and a conventional bipolar transistor (b) showing the reduction in collector/base capacitance that is obtained in the former transistor

reduction in the collector/base capacitance by as much as a factor of four. The self-alignment of the emitter to the extrinsic base allows the $p+$ extrinsic base to be brought very close to the polysilicon emitter, which dramatically reduces the extrinsic base resistance.

The essential features of the double-polysilicon, self-aligned bipolar process are illustrated in Figure 9.11. The fabrication sequence begins with the deposition of a layer of polysilicon over the top of the intrinsic base region of the transistor. This layer is then doped using a heavy boron implant, and a CVD silicon dioxide layer is deposited on top. In the completed transistor the $p+$ polysilicon forms the extrinsic base region of the transistor. At this point the oxide and polysilicon are patterned using reactive ion etching. It is important that this etch is highly anisotropic so that vertical walls are obtained at the edge of the window, as shown in Figure 9.11(a).

The critical stage of the self-aligned process is the formation of an oxide spacer on the sidewalls of the polysilicon. This is done by depositing a CVD oxide layer and then etching this back using reactive ion etching. An

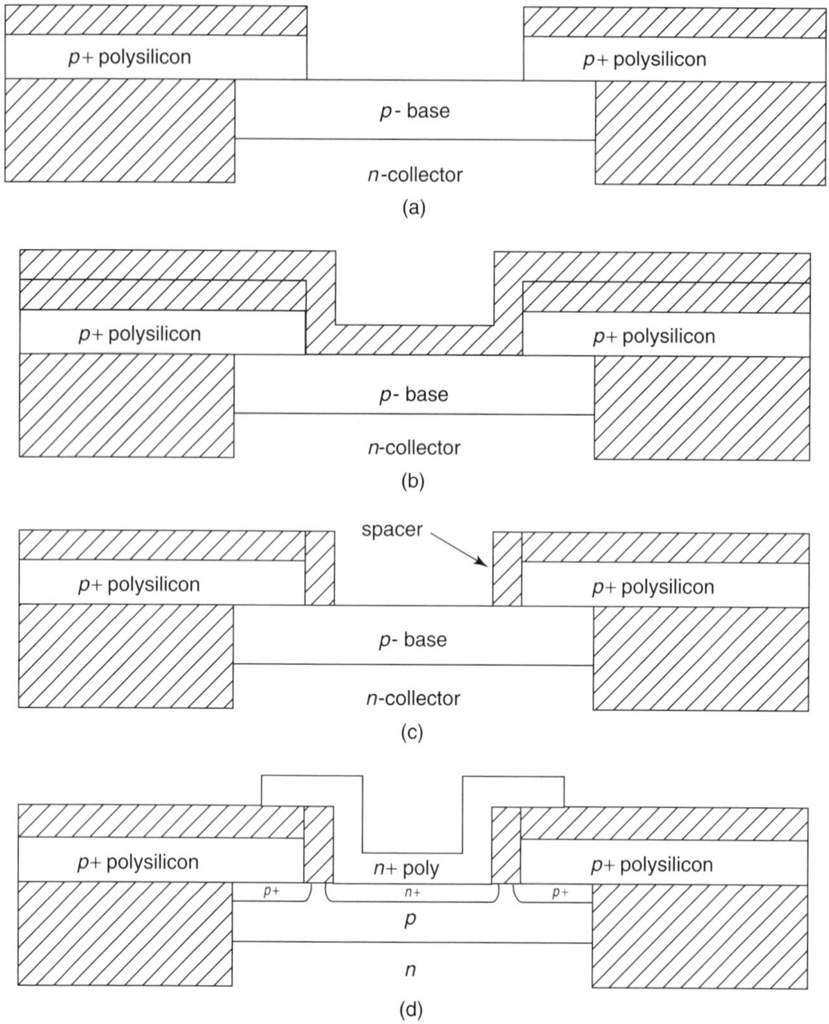

Figure 9.11 Process sequence for a double-polysilicon, self aligned bipolar process

inspection of Figure 9.11(b) indicates that the deposited oxide is thicker where it covers the step in the polysilicon. The use of an anisotropic etch to remove the oxide therefore leads to the formation of a spacer on the sidewall of the polysilicon, as shown in Figure 9.11(c). The width of the spacer is determined by the thickness of the deposited oxide layer and by the etching characteristics of the reactive ion etcher. The spacer needs to be wide enough to prevent the $p+$ extrinsic base region intersecting the $n+$ emitter region beneath the spacer (Figure 9.11(d)). If this occurs, an unwanted $p+/n+$ junction forms around the periphery of the emitter.

A large tunnelling current can flow across junctions of this type, with the result that the current gain and emitter/base breakdown voltage are significantly degraded [34]. The transistor structure is completed by forming a polysilicon emitter and annealing to drive the dopants from the two polysilicon layers into the single-crystal silicon (Figure 9.11(d)).

The self-aligned process gives a considerable reduction in the two most important electrical parameters of the bipolar transistor: the collector/base capacitance and the base resistance. It will be shown in Chapter 12 that these two parameters are the dominant components of the propagation delay in ECL logic circuits. The improvements provided by the double-polysilicon, self-aligned process therefore lead directly to a considerable improvement of circuit performance. The reduced base resistance is obtained because the $p+$ extrinsic base is self- aligned to the $n+$ polysilicon emitter. These two regions of the device are separated by the thickness of the oxide spacer, which is typically less than $0.2\,\mu m$. The extrinsic base region therefore extends right up to the edge of the active emitter, thereby providing a very low-resistance path to the base contact. A further reduction in extrinsic base resistance can be obtained by siliciding the $p+$ polysilicon layer. It should also be noted that the final emitter size is smaller than the original emitter window etched in the $p+$ polysilicon layer because of the oxide spacers on the inside of the emitter window. This allows emitters to be produced that are considerably smaller than the minimum feature size of the lithography tool.

In practice, the fabrication steps used to produce the oxide spacer are more complicated than suggested in Figure 9.11 [30,31]. Figure 9.12 shows a more realistic process sequence for the fabrication of the self-aligned polysilicon emitter. After etching of the $p+$ extrinsic base polysilicon, a thin stress relief oxide is grown, followed by the deposition of a silicon nitride layer and an undoped polysilicon layer. The undoped polysilicon layer is then anisotropically etched to create dummy polysilicon spacers at the perimeter of the emitter window, as shown in Figure 9.12(a). The silicon nitride layer acts as an etch stop for the undoped polysilicon etch. The silicon nitride layer is then etched to expose the underlying stress relief oxide in the emitter window, as shown in Figure 9.12(b). At this point, the dummy polysilicon spacer is removed using a wet etch and the stress relief oxide is wet etched to open the emitter window. The final stage in the process is the deposition of a polysilicon emitter and a high-temperature anneal to diffuse the dopants from the polysilicon layers. This process has the advantage of reducing the emitter plug effect (Section 6.8.4) compared with the process in Figure 9.11 because the insulator step height at the perimeter

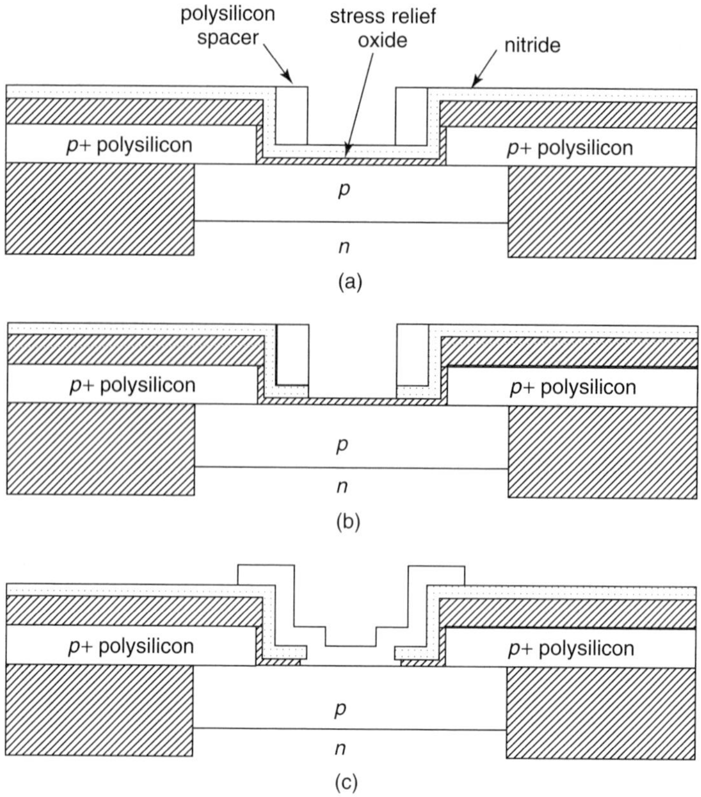

Figure 9.12 Process sequence for spacer formation in a double-polysilicon bipolar process

of the emitter is determined by the thickness of the nitride and stress relief oxide layer rather than the thickness of the $p+$ polysilicon and the overlying oxide layer.

There are two options for fabricating the intrinsic base in the double-polysilicon, self-aligned bipolar process. In the processes shown in Figures 9.11 and 9.12, the intrinsic base was fabricated before the self-aligned emitter. However, over-etching in the step shown in Figure 9.11(a) removes dopant from the intrinsic base of the transistor and hence introduces a potential problem of batch-to-batch gain variations. An alternative approach is to implant the intrinsic base after step 9.12(b), i.e. after the dummy polysilicon spacer has been produced. This approach has the advantage that the base is not dry etched after it has been implanted, and hence the gain should be very reproducible. However, the connection of the intrinsic and extrinsic bases is achieved by lateral diffusion of the extrinsic and intrinsic bases beneath the spacer. Care must therefore be taken to ensure that the intrinsic base joins up with the extrinsic base.

9.6 SINGLE-POLYSILICON BIPOLAR PROCESS

Many applications do not require the performance levels of double-polysilicon bipolar processes or SiGe HBT processes, and in such cases cost and manufacturability often determine the applicability of the technology. Single-polysilicon bipolar processes are capable of achieving reasonable performance and are considerably simpler than double-polysilicon processes. A cut-off frequency of 35 GHz has been achieved with a maximum oscillation frequency of 54 GHz [35].

In a single-polysilicon emitter, the polysilicon emitter is deposited after a window has been opened in a thin silicon dioxide layer, as illustrated in Figure 9.13. The single-polysilicon emitter is not fully self-aligned because an alignment tolerance is needed for the definition of the polysilicon around the emitter window. However, it does have some self-aligned features, in particular the self-alignment of the extrinsic base silicide to the polysilicon layer. This is achieved using a spacer on the outside of the polysilicon emitter and is identical to the self-aligned silicide (SALICIDE) process used in CMOS processes. The single-polysilicon emitter is therefore fully compatible with CMOS and hence ideal for BiCMOS technology. When compared with a double-polysilicon bipolar process, the single-polysilicon process does not have such low values of collector/base capacitance because space is needed for the contacts to the extrinsic base. Similarly, the extrinsic base resistance

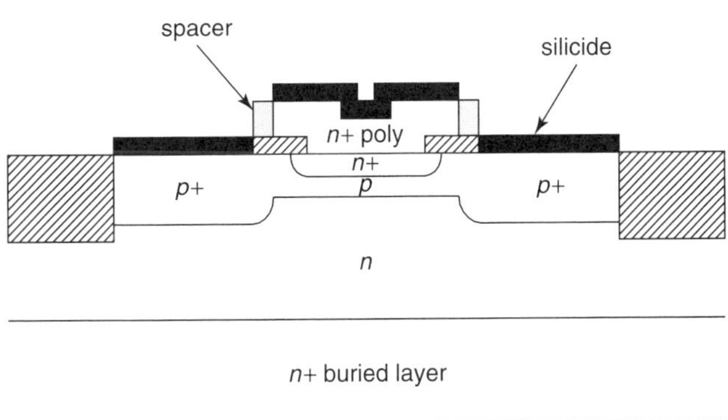

Figure 9.13 A single-polysilicon bipolar transistor

is higher because space has to be allowed for the alignment of the polysilicon to the emitter window.

9.7 BiCMOS PROCESS

CMOS is the dominant technology for the design of digital VLSI circuits because of its low power, compact layout and ease of design. However, MOS transistors have a number of disadvantages, foremost among which are a limited drive capability and a limited high-frequency performance. Bipolar transistors have a larger drive capability than MOS transistors, because the large transconductance of the bipolar transistor gives it a greater current drive per unit silicon area. BiCMOS processes allow MOS and bipolar transistors to be combined on a single chip, thereby allowing high-density MOS circuits to be combined with high-frequency and high current-drive bipolar circuits. Gate delays of bipolar [36] and BiCMOS [37] circuits degrade by approximately 50% in the presence of a large load capacitance, whereas the degradation for corresponding CMOS circuits is more than an order of magnitude. BiCMOS processes therefore offer better digital system performance than CMOS processes [38].

BiCMOS processes are also ideal for analogue and mixed-signal applications because the best features of MOS and bipolar transistors can be combined to deliver the best system performance. With analogue BiCMOS, a wide variety of different analogue and digital building blocks can be integrated onto a single chip. This system integration approach [38] enables digital functions, such as processors and memories, to be freely integrated with analogue functions, such as A-D converters, amplifiers, filters and even transducers. In this way a powerful and universal technology is created, which makes possible the integration of all types of electronic system.

The idea of merging bipolar and MOS transistors on a single chip has been around since the late 1960s [39], but little progress was made until the development of the *n*-well, silicon-gate CMOS process [40] in 1978. The *n*-well was ideal for the collector of the bipolar transistor and hence allowed for the first time the practical integration of MOS transistors with *npn* bipolar transistors to give a BiCMOS technology [41,42].

Early versions of BiCMOS technology did not use buried layer and epitaxy, and hence collector resistance was very high, which limited the current drive of the bipolar transistor. Later BiCMOS processes incorporated buried layer and epitaxy [37,38,43,44] and also included single-polysilicon emitters, as illustrated in Figure 9.14. The buried layer

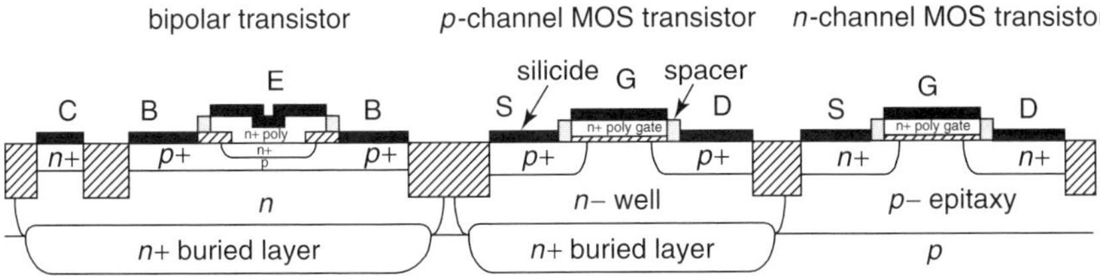

Figure 9.14 Cross-sectional view of a BiCMOS process

is also useful for the MOS transistors, since it can be introduced below the *p*-channel MOS transistor to suppress latch-up [45,46]. This buried layer reduces the series resistance of the substrate beneath the *p*-channel MOS transistor and also decreases the gain of the parasitic *pnp* bipolar transistor formed by the *p*-channel source/drain regions, the *n*-well and the *p*-substrate.

In the early BiCMOS processes, considerable effort was applied to minimizing the total number of processing steps by merging the processing of the MOS and bipolar transistors wherever possible. For example, in the process illustrated in Figure 9.14, the *p*+ source/drain implant can be used for the extrinsic base of the bipolar transistor, the *n*+ source/drain implant for the collector contact and the CMOS polysilicon gate for the polysilicon emitter of the bipolar transistor. Care must be taken when combining the polysilicon gate and polysilicon emitter because MOS transistors are very sensitive to contamination introduced between gate oxide growth and polysilicon gate deposition. For this reason, the polysilicon gate is normally deposited immediately after the gate oxide growth. In a BiCMOS process, this constraint can be met by depositing a thin polysilicon layer immediately after gate oxide growth to cap the gate oxide. The emitter window of the bipolar transistor is then opened through this capping polysilicon layer. The merging of MOS and bipolar transistor processing steps allowed BiCMOS processes to be realized using just three additional masking stages, one for the buried layer, one for the base and one for the emitter window.

In more recent BiCMOS technologies [47–51] the trend is not to merge the process steps for the MOS and bipolar transistors, but rather to add the bipolar transistor with minimum disturbance to the CMOS process. The reason for this change in emphasis is partly due to the large effort required to develop a deep sub-micron CMOS process, and partly due to the importance of time to market. Also, as extra levels of metal have been added to CMOS processes, the saving that can be achieved by eliminating a mask has decreased as a fraction of the total batch cost. Additional process steps have also been added to optimize the performance of the BiCMOS technology. Where twin tub CMOS is used, a buried *p*+ layer is often included below the *n*-channel transistor as well as a buried *n*+ layer below the *p*-channel transistor [48,49] to further reduce latch up. Deep trench isolation is also often included [48,49] to improve packing density, as discussed in Section 9.3. If high-speed performance is of paramount importance [52], a self-aligned double-polysilicon bipolar transistor can be integrated into a BiCMOS technology [49,52]. A BiCMOS technology of this type is suitable for high-speed processor and high-frequency communications applications.

9.8 COMPLEMENTARY BIPOLAR PROCESS

One of the advantages of CMOS for analogue circuit design is the availability of complementary *n*-channel and *p*-channel MOS transistors. Complementary transistors can also been obtained in bipolar technologies [51,53] by using a complementary bipolar process, which combines vertical *npn* and *pnp* bipolar transistors in a single process. Figure 9.15 illustrates a simple complementary bipolar process that incorporates single-polysilicon vertical *npn* and *pnp* bipolar transistors. Complementary bipolar processes can also be obtained with double-polysilicon bipolar transistors.

The key to the complementary bipolar process is the use of deep trench isolation to isolate the *npn* and *pnp* transistors. The *npn* transistor is fabricated in the usual way with an *n*+ buried layer and an *n*-type epitaxial layer for the collector. The *pnp* transistor requires a *p*+ buried layer, which means that an extra *n*− buried layer has to be incorporated to isolate the pnp transistor. This *n*− buried layer needs to be connected

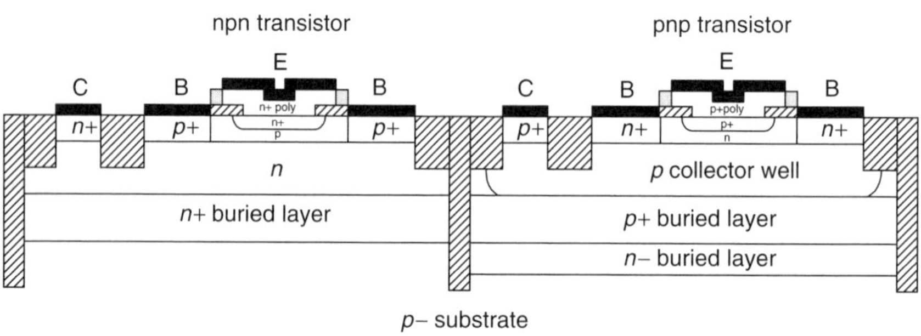

Figure 9.15 Cross-sectional view of a complementary bipolar process

to the positive supply voltage to ensure that the *pn* junction formed by the substrate and the *n*− buried layer is reverse biased. A *p*-well diffusion is used to create the collector of the *pnp* transistor by over-doping the *n*-type epitaxial layer. The process flow can be simplified by combining some of the process steps for the *npn* and *pnp* transistors. The *p*+ extrinsic base implant of the *npn* transistor can be used for the collector contact of the *pnp* transistor and vice versa for the *n*+ extrinsic base implant for the *pnp* transistor. The same undoped polysilicon layer can be used to form the polysilicon emitters of the *npn* and *pnp* transistors.

REFERENCES

[1] S.M. Sze, *VLSI Technology*, McGraw-Hill, New York (1988).

[2] A.R. Srinivasan, 'CVD epitaxial autodoping in bipolar VLSI technology', *Jnl Electrochem. Soc.*, **132**, 3005 (1985).

[3] M.W.M. Graef, B.J.H. Leunissen and H.H.C. deMoor, 'Antimony, arsenic and phosphorus autodoping in silicon epitaxy', *Jnl Electrochem. Soc.*, **132**, 1942 (1985).

[4] H.R. Chang, 'Autodoping in silicon epitaxy', *Jnl Electrochem. Soc.* **132**, 219 (1985).

[5] D. Robinson, A.A. Rozgonyi, T.E. Seidel and M.H. Read, 'Orientation and implantation effects on stacking faults during silicon buried layer processing', *Jnl Electrochem. Soc.* **128**, 926 (1981).

[6] K.V. Ravi, *Imperfections and Impurities in Semiconductor Silicon*, Wiley, Chichester (1981).

[7] P.D. Taylor, *Thyristor Design and Realization*, Wiley, Chichester (1987).

[8] J. Böck, H. Knapp, K. Aufinger, M. Wurzer, S. Boguth, R. Schreiter, T.F. Meister, M. Rest, M. Ohnemus and L. Treitinger, '12 ps implanted base silicon bipolar technology, *Proc. BCTM* (1999).

[9] S.P. Weeks, 'Pattern shift and pattern distortion during CVD epitaxy on (111) and (100) silicon', *Solid State Technol.*, **24**, 111 (1981).

[10] C.M. Drum and C.A. Clark, 'Anisotropy of macrostep motion and pattern edge displacements in silicon near (100)', *Jnl Electrochem. Soc.*, **117**, 1401 (1970).

[11] P.H. Lee, M.T. Wauk, R.S. Rosier and W.C. Benzing, 'Epitaxial pattern shift comparison in vertical, horizontal and cylindrical reactor geometries', *Jnl Electrochem. Soc.*, **124**, 1824 (1977).

[12] R.B. Herring, 'Advances in reduced pressure silicon epitaxy', *Solid State Technol.*, **22**, 75 (1979).

[13] J.A. Appels, E. Kooi, M.M. Paffen, J.J.H. Schlorje and W.H.C.G. Verkuylen, 'Local oxidation of silicon and its application in semiconductor technology', *Philips Res. Repts*, **25**, 118 (1970).

[14] K. Mikoshiba, T. Homma and K. Hamano, 'A new trench isolation technology as a replacement of LOCOS' *IEDM Technical Digest*, **578** (1984).

[15] Y. Tamaki, S. Isomae, S. Mizuo and H. Higuchi, 'Evaluation of dislocation generation at silicon nitride film edges on silicon substrates by selective oxidation', *Jnl Electrochem. Soc.*, **128**, 644 (1981).

[16] A. Hayasaka, Y. Tamaki, M. Kawamura, K. Ogiue and S. Ohwaki, 'U-groove isolation technique for high speed bipolar VLSIs', *IEDM Technical Digest*, **62** (1982).

[17] S. Suyama, T. Yachi and T. Serikawa, 'A new self-aligned well isolation technique for CMOS devices', *IEEE Trans. Electron. Devices*, **ED33**, 1672 (1986).

[18] D. Chin, S.H. Dhong and G.J. Long, 'Structural effects on a submicron trench process', *Jnl Electrochem. Soc.*, **132**, 1705 (1985).

[19] D.D. Tang, P.M. Solomon, T.H. Ning, R.D. Isaac and R.E. Burger, '1.25 micron deep groove isolated self-aligned bipolar circuits', *IEEE Jnl Solid State Circuits*, **SC17**, 925 (1982).

[20] V.J. Silvestri, 'Growth kinematics of a polysilicon trench refill process', *Jnl Electrochem. Soc.*, **133**, 2374 (1986).

[21] C.W. Teng, C. Slawinski and W. Hunter, 'Defect generation in trench isolation', *IEDM Technical Digest*, **586** (1984).

[22] K. Sagara, Y. Tamaki and M. Kawamura, 'Evolution of dislocation generation in silicon substrates by selective oxidation of U-grooves', *Jnl Electrochem. Soc.*, **134**, 500 (1987).

[23] H. Kurten, H.J. Voss, W. Kim and W.L. Engl, 'Selective low pressure silicon epitaxy for MOS and bipolar transistor application', *IEEE Trans. Electron. Devices*, **ED30**, 151 (1983).

[24] A.C. Ipri, L. Jastrzebski, J.F. Corboy and R. Metzl, 'Selective epitaxial growth for the fabrication of CMOS integrated circuits', *IEEE Trans. Electron. Devices*, **ED31**, 1741 (1984).

[25] J.F.W. Schiz, A.C. Lamb, F. Cristiano, J.M. Bonar, P. Ashburn, S. Hall and P.L.F. Hemment, 'Leakage current mechanisms in SiGe heterojunction bipolar transistors fabricated using selective and non selective epitaxy', *IEEE Trans. Electron. Devices*, **48**, 2492 (2001).

[26] A. Ishitani, H. Kitajima, N. Endo and N. Kasai, 'Facet formation in selective silicon epitaxial growth', *Japan Jnl Appl. Phys.*, **24**, 1267 (1985).

[27] C.I. Drowley and M.L. Hammond, 'Conditions for uniform selective epitaxial growth', *Solid State Technology*, **33**, 135 (1990).

[28] T.H. Ning, R.D. Isaac, P.M. Solomon, D.D. Tang, H.N. Yu, G.C. Feth and S.K. Wiedmann, 'Self-aligned bipolar transistors for high performance and low power delay VLSI', *IEEE Trans. Electron. Devices*, **ED28**, 1010 (1981).

[29] M. Suzuki, K. Hagimoto, H. Ichino and S. Konaka, 'A 9 GHz frequency divider using silicon bipolar super self-aligned process technology', *IEEE Electron. Device Lett.*, **EDL6**, 181 (1985).

[30] T. Sakai, Y. Yamamoto, Y. Kobayashi, K. Kawarada, Y. Inabe, T. Hayashi and H. Miyanaga, 'A 3 ns 1 Kbit RAM using super self-aligned process technology', *IEEE Jnl Solid State Circuits*, **SC16**, 424 (1981).

[31] T. Sakai, S. Konaka, Y. Kobayashi, M. Suzuki and Y. Kawai, 'Gigabit logic bipolar technology: advanced super self-aligned process technology', *Electron. Lett.*, **19**, 283 (1983).

[32] M. Nanba, T. Uchino, M. Kondo, T. Nakamura, T. Kobayashi, Y. Tamaki and M. Tanabe; 'A 64 GHz f_T and 3.6 V BV_{CEO} Si bipolar transistor using in-situ phosphorus doped and large grained polysilicon emitter contacts', *IEEE Trans. Electron. Devices*, **ED40**, 1563 (1993).

[33] M. Ugajin, J. Kodate, Y. Kobayashi, S. Konaka and T. Sakai; 'Very high f_T and $f_{\max}$ silicon bipolar transistors using ultra high performance, super self-aligned process technology for low energy and ultra-high speed LSI's', *Proc. IEDM*, 735 (1995).

[34] A. Cuthbertson and P. Ashburn, 'Self-aligned transistors with polysilicon emitters for bipolar VLSI', *IEEE Jnl Solid State Circuits*, **SC20**, 162 (1985).

[35] S. Niel, O. Rozeau, L. Ailloud, C. Hernandez, P. Llinares, M. Guillermet, J. Kirtsch, A. Monroy, J. de Pontcharra, G. Auvert, B. Blanchard, M. Mouis, G. Vincent and A. Chantre, 'A 54 GHz $f_{\max}$ implanted base 0.35 μm single polysilicon bipolar technology', *IEDM Technical Digest*, 807 (1997).

[36] W. Fang, A. Brunnschweiler and P. Ashburn, 'An accurate analytical BiCMOS delay expression and its application to optimising high-speed BiCMOS circuits', *IEEE Jnl Solid State Circuits*, **27**, 191 (1992).

[37] A.R. Alvarez, P. Meller and B. Tien, '2 micron merged bipolar-CMOS technology', *IEDM Technical Digest*, 761 (1984).

[38] H. Higuchi, G. Kitsukawa, T. Ikeda, Y. Nishio, N. Sasaki and K. Ogiue, 'Performance and structures of scaled-down bipolar devices merged with CMOSFETS', *IEDM Technical Digest*, 694 (1984).

[39] H.C. Lin, J.C. Ho, R.R. Iyer and K. Kwong, 'Complementary MOS-bipolar structure', *IEEE Trans. Electron. Devices*, **ED16**, 945 (1968).

[40] J. Schneider, G. Zimmer and B. Hoefflinger, 'A compatible NMOS, CMOS metal gate process', *IEEE Trans. Electron. Devices*, **ED25**, 832 (1978).

[41] G. Zimmer, B. Hoefflinger and J. Schneider, 'A fully implanted NMOS, CMOS, bipolar technology for VLSI of analog-digital systems', *IEEE Jnl Solid State Circuits*, **SC14**, 312 (1979).

[42] P.M. Zeitzoff, C.N. Anagnostopoulos, K.Y. Wong and B.P. Brandt, 'An isolated vertical *npn* transistor in an *n*-well CMOS process', *IEEE Jnl Solid State Circuits*, **SC20**, 489 (1985).

[43] H. Momose, H. Shibata, S. Saitoh, J. Miyamoto, K. Kanzaki and S. Hohyama, '*1.0 p.m* *n*-well CMOS/bipolar technology', *IEEE Trans. Electron. Devices*, **ED32**, 217 (1985).

[44] T. Ikeda, T. Nagano, N. Momma, K. Miyata, H. Higuchi, M. Odaka and K. Ogiue, 'Advanced BICMOS technology for high-speed VLSI', *IEDM Technical Digest*, 408 (1986).

[45] J. Agraz-Guerera, R.A. Ashton, W.J. Bertram, R.C. Melin, R.C. Sun and J.T. Clemens, 'Twin-tub III: a third generation CMOS technology', *IEDM Technical Digest*, 63 (1984).

[46] R.J. Smith, G. Sery, J. McCollum, J. Orton, B. Mantha, J. Smudski, T. Chi, S. Smith, J.P. Dishaw and K. Kokkonen, 'A double layer metal CHMOS III technology', *IEDM Technical Digest*, 56 (1984).

[47] H. Tian, A. Perera, C. Subramanian, D. Pham, J. Damiano, J. Scott, T. McNelly, R. Zaman and J. Hayden, 'Bipolar process integration for a 0.25 μm BiCMOS SRAM technology using shallow trench isolation', *Proc. BCTM*, 76 (1997).

[48] H. Nii, T. Yoshino, K. Inoh, N. Itoh, H. Nakajima, H. Sugaya, H. Naruse, Y. Katsumata and H. Iwai, '0.3 μm BiCMOS technology for mixed analog/digital application systems', *Proc. BCTM*, 68 (1997).

[49] Y. Kinoshita, H. Suzuki, S. Nakamura, M. Fukaishi, A. Tajima, Y. Suemura, T. Itani, H. Miyamoto, H. Fujii, M. Yotssuyanagi, N. Henmi and T. Yamazaki, 'An advanced 0.25 μm BiCMOS process integration technology for multi-GHz communication LSIs', *Proc. BCTM*, 72 (1997).

[50] D. Doyle, K. Moloney, D. Rohan, S. Feindt, K. Kattmann, C. McLoughlin, P. Meehan, S. Healy, J. Prendergast and M. O'Neill, '0.6 μm BiCMOS technology for rf and high speed converter applications', *Proc. BCTM*, 64 (1997).

[51] Y. Yoshida, H. Suzuki, Y. Kinoshita, H. Fujii and T. Yamazaki, 'An RF BiCMOS process using high f_{SR} spiral inductor with premetal deep trenches and a dual recessed bipolar collector sink', *IEDM Technical Digest*, 213 (1998).

[52] T. Hashimoto, T. Kikuchi, K. Watanabe, N. Ohashi, T. Saito, H. Yamaguchi, S. Wada, N. Natsuaki, M. Kondo, S. Kondo, Y. Homma, N. Owada and T. Ikeda, 'A 0.2 μm bipolar-CMOS technology on bonded SOI with copper metallisation for ultra-high speed processors', *IEDM Technical Digest*, 209 (1998).

[53] R. Bashir *et al.*, 'A complementary bipolar technology family with a vertically integrated pnp for high frequency analog applications', *IEEE Trans. Electron. Devices*, 48, 2525 (2001).

10

Silicon-Germanium Heterojunction Bipolar Technology

10.1 INTRODUCTION

In the 1990s a revolution in silicon technology occurred, as the Silicon-Germanium Heterojunction Bipolar Transistor (SiGe HBT) emerged from research labs around the world [1,2]. Previously, heterojunction devices had only been available in compound semiconductor technologies, and as a consequence silicon technology could not fully compete in high-frequency applications. The emergence of the SiGe HBT as a viable production device turned silicon into a heterojunction technology and in doing so extended its capability to much higher frequencies. State of the art SiGe HBTs currently have values of f_T of 350 GHz and values of $f_{\max}$ of 270 GHz [3], which implies the operation of communication systems at >100 Gbit/s. Furthermore, research is underway on SiGe heterojunction MOSFETs, which in turn will revolutionize the future of CMOS technology. It is clear therefore that SiGe heterojunction devices will have a big impact on the future of silicon technology.

Two main approaches have been used for the fabrication of SiGe HBTs, as illustrated in Figure 10.1. The first uses differential epitaxy [4,5], as illustrated in Figure 10.1(a), and the second selective epitaxy [6,7], as illustrated in Figure 10.1(b). For the differential epitaxy process, the $p+$ SiGe base layer is grown after oxide isolation formation, so single-crystal material is formed where the silicon collector

SiGe Heterojunction Bipolar Transistors Peter Ashburn
 ISBN: 0-470-84838-3

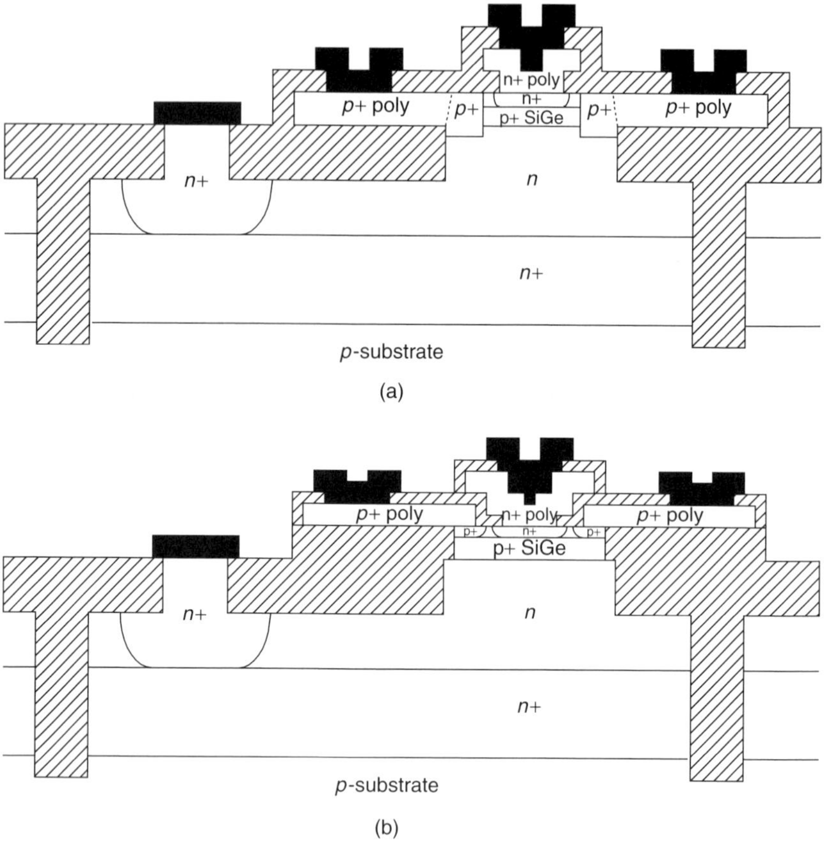

Figure 10.1 Cross-sectional views of SiGe HBTs produced using (a) differential epitaxy and (b) selective epitaxy

is exposed and polycrystalline material over the oxide isolation. This polycrystalline layer is used to form the extrinsic base contact. For the selective epitaxy process, the $p+$ polysilicon extrinsic base is formed before epitaxy, and the SiGe base grown using selective epitaxy. This approach is a variant of the double-polysilicon self-aligned bipolar process. In the following sections these two approaches will be described in more detail, along with variants for SiGe:C HBT processes. The alternative approach of using germanium implantation to fabricate SiGe HBTs will also be considered. Finally, as SiGe HBTs are extensively used in high-frequency communications circuits, the methods of fabricating the passive components (resistors, capacitors and inductors) will be described.

10.2 DIFFERENTIAL EPITAXY SILICON-GERMANIUM HBT PROCESS

Figure 10.2 summarizes a simplified process sequence for the fabrication of a SiGe HBT using differential epitaxy [5,8]. The starting point for the sequence is after the formation of deep trench isolation (not shown) and shallow trench isolation. A *p*+ SiGe base and *p* Si cap is then grown non-selectively to give single-crystal material where the silicon collector is exposed and polycrystalline material over the shallow trench isolation. The interface between the polycrystalline and single-crystal material angles diagonally upwards from the top corner of the shallow trench isolation, as illustrated in Figure 10.2(b).

Figure 10.2(c) shows the SiGe HBT at extrinsic base formation. This stage is reached by depositing a thin oxide layer, opening an emitter window, depositing an *n*+ polysilicon layer, and etching to create the polysilicon emitter. At this point, an extrinsic base implant is performed to form a heavily doped *p*+ extrinsic base region adjacent to the polysilicon emitter and to dope the polycrystalline layer heavily *p*-type. It should be noted that the extrinsic base implant penetrates into the SiGe layer and hence the point defects created will give rise to transient enhanced diffusion during later annealing, as discussed in Section 8.7.2. The use of SiGe:C is therefore an attractive option for this process, as discussed in Section 8.7.3. Following the extrinsic base implantation, an anneal is carried out to anneal the implantation damage and diffuse the arsenic from the polysilicon emitter into the *p*-type silicon cap. The arsenic should diffuse to a sufficient depth to over-dope the *p*-type silicon cap and penetrate into the SiGe base. If a graded germanium profile is used, like that in Section 8.6, the exact position of the emitter profile with respect to the germanium profile will not be too critical, because the majority of the germanium is at the collector end of the base.

Figure 10.2(d) shows the completed SiGe HBT. It can be seen that this is a double-polysilicon process and hence has the advantage of low collector/base capacitance, as discussed in Section 9.5. The *p*+ polysilicon can be silicided to reduce the contribution of the polysilicon layer to the extrinsic base resistance. The largest component of the extrinsic base resistance is then due to the series resistance of the *p*-type silicon cap beneath the overhanging polysilicon emitter. The extrinsic base is self-aligned to the edge of the overhanging polysilicon emitter, but an alignment tolerance is still needed between the emitter window and the polysilicon, as discussed in Section 9.6. This alignment tolerance

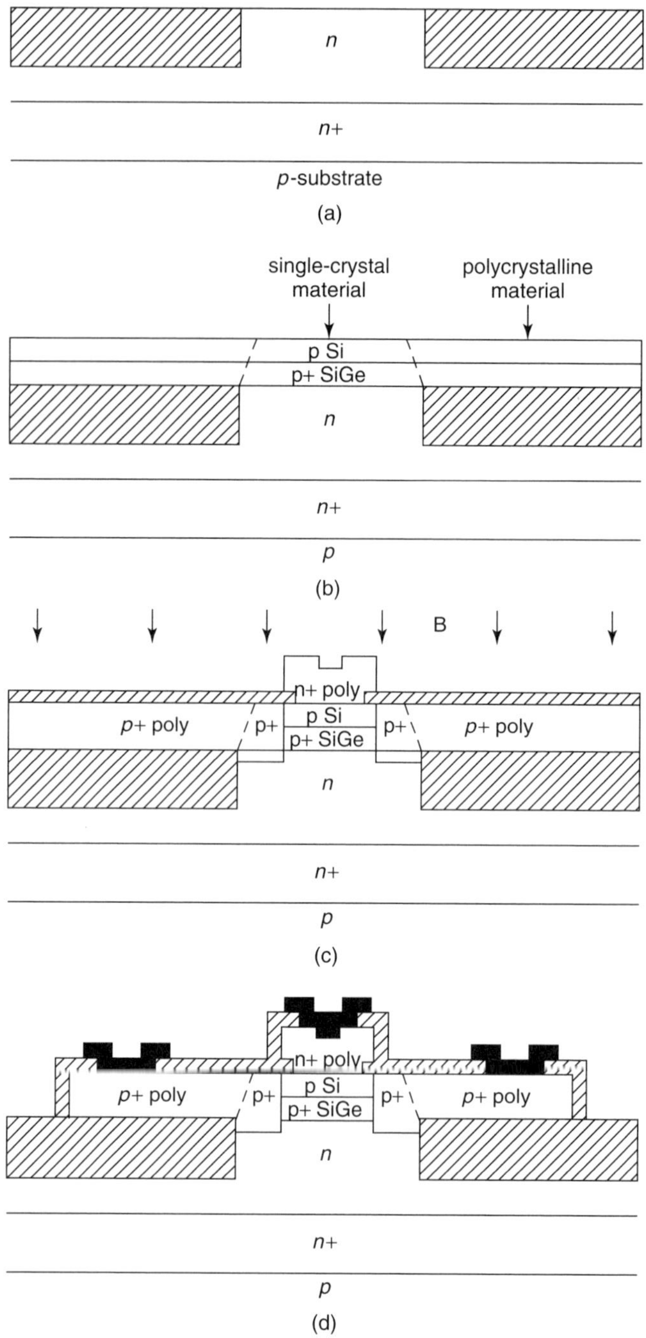

Figure 10.2 Simplified process sequence for a SiGe HBT fabricated using differential epitaxy

will have a strong influence on the extrinsic base resistance. An example of a self-aligned emitter will be given in Section 10.2.2, which reduces this component of extrinsic base resistance.

10.2.1 Polysilicon Nucleation Layer

In the differential epitaxy SiGe HBT process, the nucleation of polycrystalline SiGe on the field oxide is sometimes poor because the growth processes used are often slightly selective in the initial stage of the growth. In this situation, a polysilicon nucleation layer can be included on top of the shallow trench isolation, as illustrated in Figure 10.3. After shallow trench isolation, a thin oxide layer is grown and a thin polysilicon nucleation layer deposited. The polysilicon nucleation layer is then etched using the underlying oxide layer as an etch stop, and the oxide layer wet etched to give the structure shown in Figure 10.3(a). Epitaxial growth of the $p+$ SiGe and p-Si cap is then performed, as illustrated in Figure 10.3(b). The thickness of the polysilicon nucleation

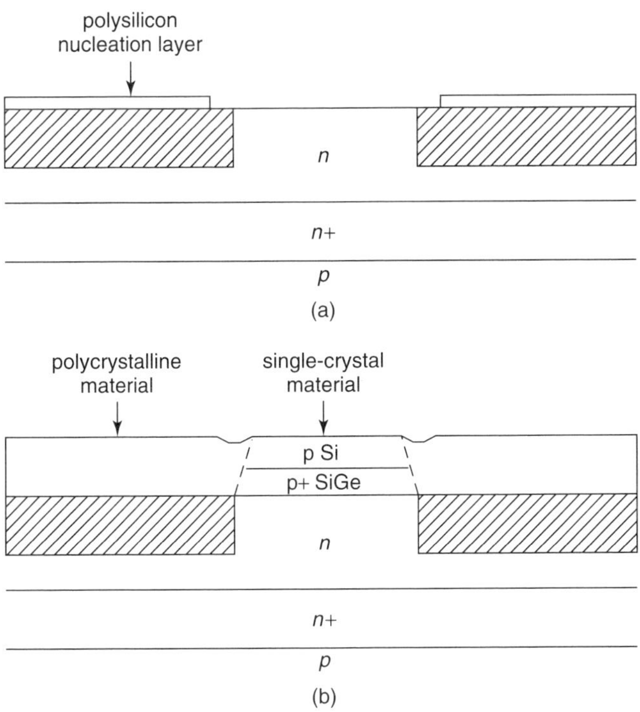

Figure 10.3 Use of a polysilicon nucleation layer to increase the thickness of the polycrystalline material on the field oxide

layer can be chosen to compensate for any difference in growth rate between the single-crystal and polycrystalline material, so that a planar surface is obtained after growth.

The polysilicon nucleation layer also serves a number of other useful functions. In a BiCMOS process, a gate oxide layer would be grown after shallow trench isolation, and a thin polysilicon layer deposited to protect the gate oxide from contamination during emitter processing, as discussed in Section 9.7. In a SiGe HBT, the gate-protect polysilicon layer can serve the dual function of protecting the MOS transistors during the epitaxial growth and acting as a nucleation layer for the extrinsic base polysilicon. The polysilicon nucleation layer is also useful during pre-epitaxy cleaning, particularly if the hydrogen passivation approach is used, as discussed in Section 7.5.2. Hydrogen passivation gives a hydrophobic surface after the clean, which is very difficult to detect when the majority of the wafer is covered by silicon dioxide. If a polysilicon nucleation layer is included, the majority of the wafer is covered by either polysilicon or exposed single-crystal silicon, and hence the hydrophobic surface is very easy to detect. A final advantage of the polysilicon nucleation layer is that the growth conditions are almost identical to those for a blanket silicon wafer, since the majority of the wafer surface is covered with silicon. In this situation, the development of the growth process is relatively straightforward.

10.2.2 Self-aligned Emitter for the Differential Epitaxy HBT

In the process discussed in Figure 10.2, the emitter was only quasi self-aligned, since an alignment tolerance was needed between the emitter window and the $n+$ polysilicon. This approach has the advantage of simplicity, but the disadvantage that the single-crystal emitter is not symmetrically located between the extrinsic base regions, because the separation on either side depends on the accuracy of the alignment. This uncertainty in the location of the single-crystal emitter with respect to the extrinsic base can give rise to some variability in the value of extrinsic base resistance.

Self-aligned emitters have been developed for application in differential epitaxy SiGe HBT technology [5], but at the cost of increased complexity. Figure 10.4 illustrates the process sequence. Following base and emitter epitaxy, a thin stress relief oxide layer is deposited, followed by a silicon nitride layer and a polysilicon conversion layer. A silicon nitride layer and a thick silicon dioxide layer are then deposited and patterned to create a dummy emitter, as illustrated in Figure 10.4(a).

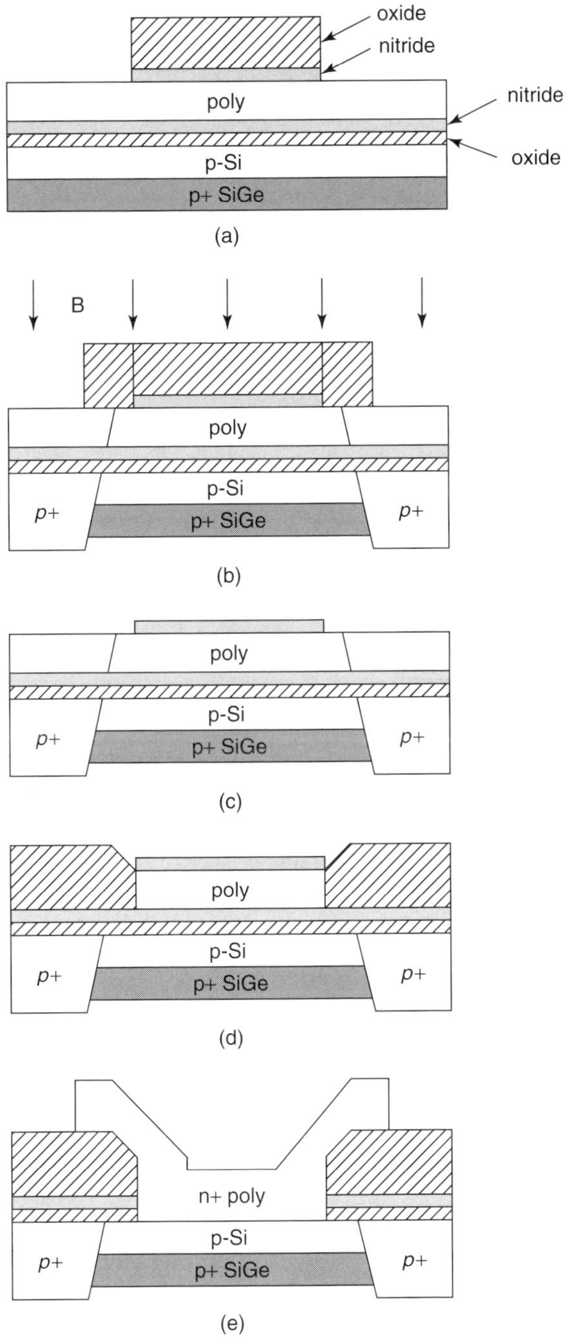

Figure 10.4 Process sequence for the fabrication of a self-aligned emitter in a differential epitaxy SiGe HBT (reprinted with permission from [5])

The extrinsic base fabrication is begun by depositing a silicon dioxide layer and then using an anisotropic etch to give oxide spacers on the side of the dummy emitter, as shown in Figure 10.4(b). An extrinsic base implant is then performed, which is self-aligned to the dummy emitter. The oxide spacers ensure that the extrinsic base is separated from the polysilicon emitter. Following the extrinsic base implant, the top oxide layer is removed to leave the silicon nitride layer on top of the conversion polysilicon layer, as shown in Figure 10.4(c).

The self-aligned emitter is formed by locally oxidizing the polysilicon conversion layer, using the nitride layer to prevent oxidation over the emitter, as shown in Figure 10.4(d). High-pressure oxidation is preferred for this step, since it can be done at a low temperature, thereby avoiding diffusion of the boron in the SiGe base. Following oxidation, the nitride layer and remaining polysilicon conversion layer can be removed and the emitter window opened in the nitride and stress relief oxide using a wet etch. The final step is the creation of a polysilicon emitter to give the structure shown in Figure 10.4(e).

10.3 SELECTIVE EPITAXY SILICON-GERMANIUM HBT PROCESS

The differential epitaxy process has the disadvantage that the SiGe base is implanted when the extrinsic base is formed. The damage created gives rise to transient enhanced diffusion of the boron in the SiGe base during subsequent high-temperature anneals, as discussed in Section 8.7.2. This problem can be avoided if the extrinsic base is fabricated before the SiGe epitaxy, so that the SiGe layer does not need to be implanted after epitaxy. This can be achieved if selective epitaxy is used to grow the SiGe base [9–12].

Figure 10.5 illustrates the process sequence for the selective epitaxy SiGe HBT process. The process begins with the growth of a thermal silicon dioxide layer, the deposition of a polysilicon layer and doping with a high dose boron implant to create an extrinsic base as shown in Figure 10.5(a). A silicon nitride layer is then deposited and an emitter window etched in the nitride and $p+$ polysilicon layers to give the structure shown in Figure 10.5(b). A silicon nitride spacer is formed on the sidewalls of the $p+$ polysilicon extrinsic base by nitride deposition and anisotropic etch, and the bottom oxide layer is wet etched laterally to expose the bottom face of the $p+$ polysilicon extrinsic base, as shown in Figure 10.5(c). During epitaxy, single-crystal SiGe grows on the silicon collector and polycrystalline SiGe on the exposed $p+$ polysilicon

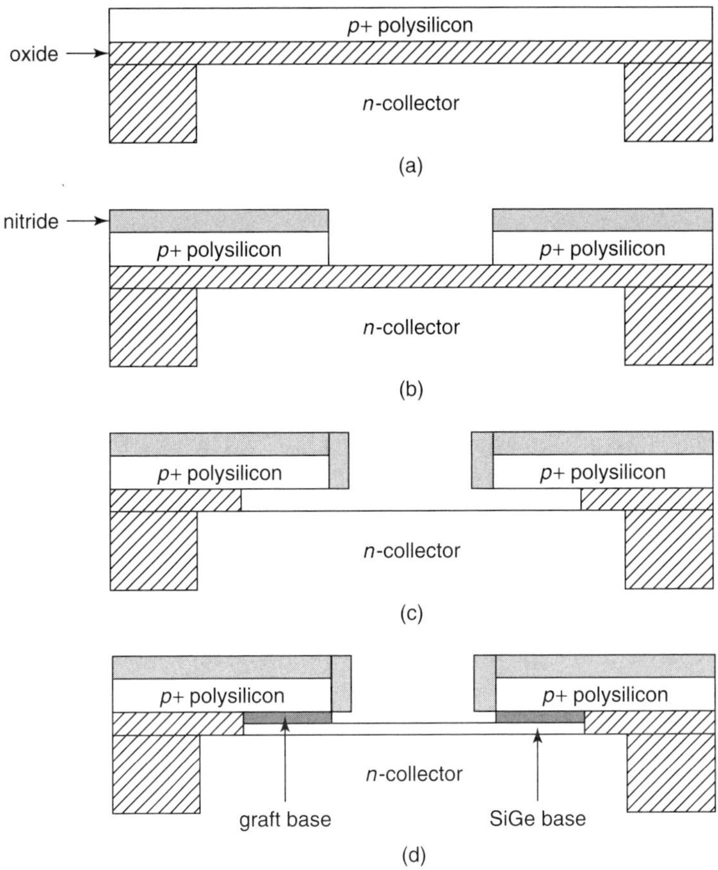

Figure 10.5 Simplified process sequence for a SiGe HBT fabricated using selective epitaxy

extrinsic base to create a graft base, as shown in Figure 10.5(d). Once the graft base has joined up to the SiGe base, the silicon emitter cap layer can be grown.

The major disadvantage of this process is the difficulty in controlling the selective epitaxy growth in a production environment. As discussed in Section 7.7.1, selective epitaxy has the two major disadvantages of facet formation [13] and loading effects [14]. In this SiGe HBT process, loading effects will cause the SiGe thickness to vary when the emitter window size is varied. Different geometry transistors will therefore have different basewidths and hence different values of gain and cut-off frequency. Facet formation during selective epitaxy will tend to lead to the formation of voids in the corners of the emitter window adjacent to the graft base. Such voids will make it difficult to ensure that the graft base is properly joined to the SiGe base. More sophisticated graft base processes have been devised [11] to reduce void formation and hence improve the reliability of the connection between the graft base and the SiGe base.

10.4 SILICON-GERMANIUM-CARBON HBT PROCESS

As discussed above, the main disadvantage of the differential epitaxy SiGe HBT process is the transient enhanced boron diffusion in the SiGe base that occurs as a result of the damage from the extrinsic base implant. However, the introduction of carbon into a SiGe base has been shown to dramatically suppress transient enhanced diffusion of boron [15–18], as discussed in Section 8.7.3. It is therefore clear that the introduction of carbon into the differential epitaxy SiGe HBT process would be expected to deliver major improvements in transistor performance.

Several groups have investigated the use of carbon in differential epitaxy SiGe HBT processes and demonstrated its benefits [19–21]. A carbon content of around 0.1% has been shown to suppress both thermal and transient enhanced boron diffusion for boron doping levels up to $3 \times 10^{19}\,\mathrm{cm}^{-3}$ [20]. This has allowed SiGe:C HBTs to be produced using differential epitaxy processes with values of f_T of 350 GHz [21]. The introduction of carbon into the differential epitaxy SiGe HBT process therefore eliminates transient enhanced diffusion of boron and hence provides resilience against implantation steps and associated high-temperature anneals. With SiGe:C HBTs it is therefore possible to implant both the extrinsic base and the selective implanted collector after growth of the SiGe layer [20].

As discussed in Section 8.7.3, the introduction of carbon into SiGe HBTs suppresses both the thermal diffusion of boron and the transient enhanced diffusion of boron. This suppression of the thermal diffusion of boron implies that introduction of carbon would also be beneficial to SiGe HBTs produced using the selective epitaxy process. Several groups have incorporated carbon into selective epitaxy SiGe HBT processes to assess the improvement in performance that can be achieved [22,23]. Böck *et al.* [22] introduced a carbon concentration of $6 \times 10^{19}\,\mathrm{cm}^{-3}$ into selective epitaxy SiGe HBTs and showed that the carbon suppressed undesirable boron diffusion for a boron base doping of $2 \times 10^{19}\,\mathrm{cm}^{-3}$. Oda *et al.* [23] investigated carbon contents in the range 0.2–0.4% and showed that a 10 nm thick boron doped layer could be produced with a boron doping concentration of $4\times10^{19}\,cm^{-3}$. Both the f_T and the f_{max} were found to increase with carbon content. The carbon also reduced facet formation and decreased the standard deviation of collector current variations across the wafer due to improved thermal stability.

10.5 SILICON-GERMANIUM HBT PROCESS USING GERMANIUM IMPLANTATION

While the majority of SiGe HBTs have been produced using epitaxy, some work has been done on the use of germanium implantation to produce SiGe HBTs [24–27]. Germanium implantation has the advantage of being fully compatible with the self-aligned double polysilicon bipolar process because it allows SiGe layers to be easily produced in selected areas of the silicon wafer. This advantage makes germanium implantation attractive as a low cost route to the realization of a SiGe HBT technology. The main disadvantage of germanium implanted SiGe HBTs is that very narrow basewidths cannot be achieved since the base is produced using boron implantation. For this reason, germanium implanted SiGe HBTs have been limited to values of f_T and f_{max} of around 75 GHz [28].

Figure 10.6 shows a schematic cross-section of the germanium implantation, which occurs immediately before the intrinsic base implantation [24]. The dose of the germanium implant is typically around $3\times10^{16}\,cm^{-2}$, which is high enough to amorphize the silicon. The boron for the intrinsic base is then implanted into the amorphized silicon, which has the advantage of eliminating the channelling tail on the implanted boron profile. Furthermore, the presence of the germanium in the base reduces boron diffusion during the later implantation anneal, since boron diffusion in SiGe is slower than that in Si, as discussed in Section 8.7.2. These two factors give significantly narrower basewidths

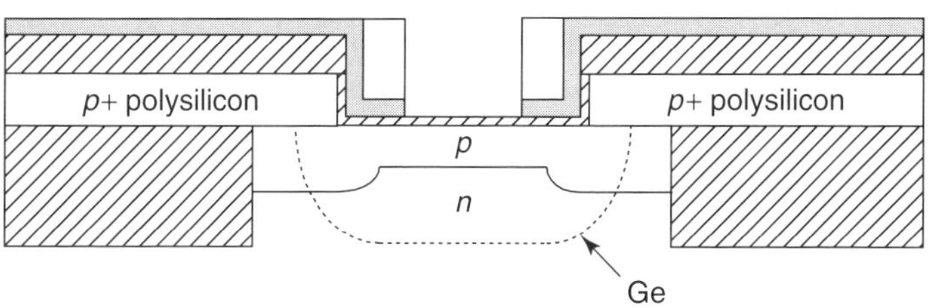

Figure 10.6 Germanium implantation in a Ge implanted SiGe HBT process

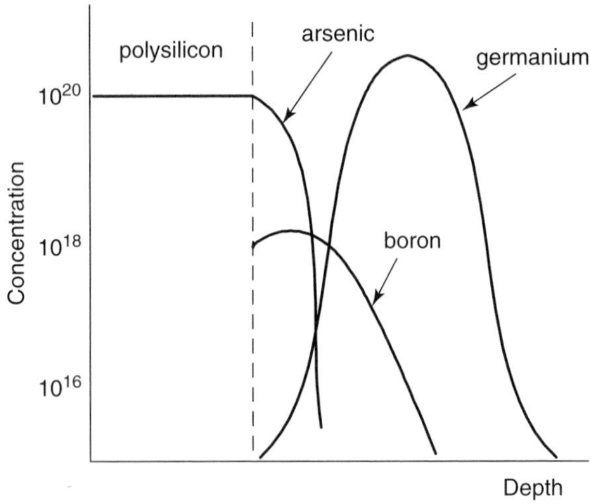

Figure 10.7 Typical germanium, boron and arsenic profiles in a germanium implanted SiGe HBT

in germanium implanted SiGe HBTs than in comparable silicon bipolar transistors.

Figure 10.7 shows typical germanium and boron profiles for the germanium implanted SiGe HBT. The energy of the germanium implant is chosen so that the peak of the germanium profile occurs close to the collector/base junction. In this situation, the germanium concentration rises across the base and gives a graded germanium profile analogous to that used in epitaxial SiGe HBTs. As discussed in Section 8.6, this graded germanium profile gives a built-in electric field that accelerates electrons across the base and hence increases the f_T of the SiGe HBT.

The main issue in germanium implanted SiGe HBTs is the control of defects resulting from the germanium implant. Transmission electron microscopy has shown that good epitaxial quality can be achieved in the re-crystallized SiGe layer and that strained SiGe can be produced [29]. This is done using a two-stage thermal anneal to achieve solid phase epitaxial regrowth of the SiGe layer and annihilation of electrically active defects [29]. The first stage is a re-crystallization anneal at a temperature between 500 and 600°C and the second stage is a higher temperature anneal at ≈850°C to annihilate remaining defects. A band of defects is invariably present at the end of range of the germanium implant, but these do not have a detrimental effect on the transistor performance since they are located in the collector of the transistor [24].

10.6 RADIO FREQUENCY SILICON-GERMANIUM BiCMOS PROCESS

While the digital BiCMOS process in Section 9.7 needed little more than MOS and bipolar transistors, analogue BiCMOS processes require a wide range of additional components such as resistors, capacitors, diodes and *pnp* bipolar transistors. RF BiCMOS processes require, in addition, inductors and varactor diodes. While some of these passive components can be fabricated without any additional process steps, others require extra processing.

Resistors can easily be produced without any additional processing by using the series resistance of the various layers that comprise the bipolar and MOS transistors in a BiCMOS technology. Implanted *n*+ and *p*+ resistors can be produced using the *n*+ and *p*+ source/drain implants of the *n*-channel and *p*-channel MOS transistors respectively or the *p*+ extrinsic base implant of the bipolar transistor. Sheet resistances of typically 70 and 100 Ω/sq, respectively, can be obtained in this way. A higher value of sheet resistance can be obtained by using a lower dose boron implant analogous to that used to create the base of a silicon bipolar transistor. A sheet resistance of around 1–2 kΩ/sq can be obtained in this way. Control of the absolute value of resistance can be obtained to about ±10%, while matching between adjacent resistors is possible to a much better tolerance of typically ±0.1%. Figure 10.8 shows a cross-section of a *p*+ resistor formed in an *n*-well. It should be noted that the resistor has to be electrically isolated from the rest of the circuit in the same way as a bipolar transistor, which is done by connecting the *n*-well to the positive supply voltage. The disadvantages of implanted resistors are that high-value resistors consume considerable silicon area and that they have a high parasitic capacitance due to the presence of the depletion region.

Polysilicon resistors are an alternative to implanted resistors and can be produced using the *n*+ and *p*+ polysilicon of the *n*-channel and *p*-channel MOS transistors, respectively, and the *n*+ polysilicon of the bipolar transistor polysilicon emitter. Polysilicon resistors are formed

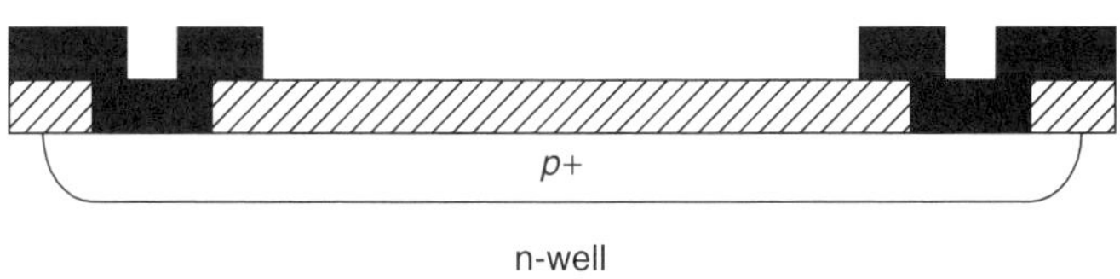

Figure 10.8 Cross-section of a *p*+ implanted resistor

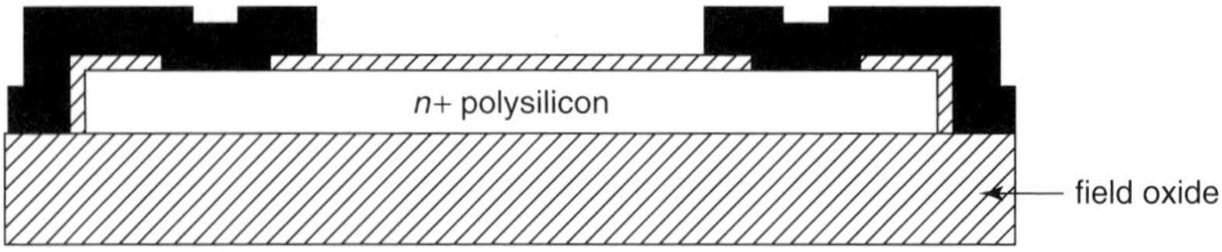

Figure 10.9 Cross-section of an *n*+ polysilicon resistor

on top of the thick field oxide, as shown in Figure 10.9, and hence have a lower parasitic capacitance than implanted resistors. They also do not directly consume any silicon area, as they are located above the silicon substrate. A further advantage of a polysilicon resistor over an implanted resistor is that they do not require any isolation connections, as they are isolated by the underlying field oxide. Sheet resistances of 200–300 Ω/sq are typically obtained for polysilicon resistors and control of the absolute value of resistance can be achieved to about ±20%. This poor tolerance is the main disadvantage of polysilicon resistors, and is caused by uncertainties in the series resistance of the polysilicon due to the presence of grain boundaries. Higher value polysilicon resistors can be obtained if a lower dose implant is used to dope the polysilicon. Sheet resistances of 5 kΩ/sq and more can be obtained in this way, but control of the absolute value of resistance is difficult because of the strong effect that the grain boundaries have on the resistance in lightly doped polysilicon.

MOS capacitors can easily be produced using a thin silicon dioxide layer, as illustrated in Figure 10.10. This is essentially a parallel plate capacitor, the bottom plate being provided by a low-resistance *n*+ layer. If the collector sink region over the *n*+ buried layer is used as the bottom plate, as shown in Figure 10.10, a very low series resistance is obtained and hence a Q of around 20 can be achieved at 2 GHz [30]. The capacitance per unit area depends on the oxide thickness, which

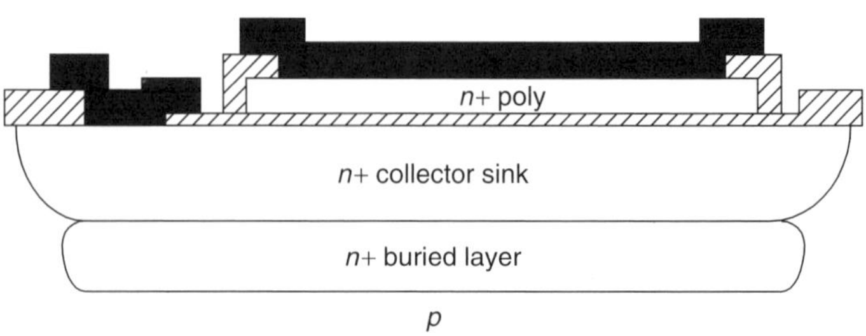

Figure 10.10 Cross-section of a MOS capacitor

is generally around 10 nm. This gives a capacitance per unit area of around 3 fF/μm^2, which can be achieved with a tolerance of ±15%. The low capacitance per unit area of this type of capacitor demonstrates that large-value capacitors are extremely expensive in terms of silicon area.

An alternative to the MOS capacitor is the Metal-Insulator-Metal (MIM) capacitor, which is made by sandwiching an insulator between two levels of metal in a multi-level metal technology. Deposited silicon dioxide is generally used as the insulator because it has a relatively high dielectric constant and can be deposited at a low temperature (below the melting point of aluminium) using plasma enhanced chemical vapour deposition. The use of two metal plates for the MIM capacitor means that series resistance is extremely low, and the large thickness of insulator between the metal plates and ground gives low parasitic capacitance. Consequently Q values of 70–80 at 2 GHz can be obtained, which makes the MIM capacitor the preferred choice for RF circuits. The capacitance per unit area achievable with a MIM capacitor depends on the oxide thickness, which is determined by the requirement for good reliability. An oxide thickness of around 50 nm is typically used, which gives a capacitance per unit area of approximately 0.7 fF/μm^2 [31].

The realization of high Q inductors in RF BiCMOS technologies is difficult because of the compromises that have to be made between the inductor performance and the performances of the MOS and bipolar transistors. Inductors are generally fabricated by realizing a metal spiral in the top level of metallization, as illustrated in Figure 10.11. Contact to the centre of the spiral is made to a lower level of metal through a via. The Q of the inductor is limited by the series resistance of the metal, which is typically 10–100 mΩ/sq and by the parasitic capacitance to the silicon substrate, which is determined by the oxide thickness. Decreasing the series resistance of the inductor metal has a big effect on the Q of the inductor. This can be done by increasing the thickness of the inductor metal layer and/or by using a low-resistance metal such as copper. Decreasing the parasitic capacitance of the inductor generally has a smaller effect on the Q of the inductor than decreasing the series resistance. Values of Q between 15 and 20 at 2 GHz can be achieved by using a 4–5 μm thick aluminium layer and a 3 μm thick oxide layer between the inductor metal layer and the previous metal layer [31]. Values of inductance depend on the number of turns and are typically in the range 1–10 nH.

The substrate resistivity is important in RF BiCMOS technologies because interactions between the inductor electric and magnetic fields and the substrate can lead to parasitic substrate currents [31]. Parasitic

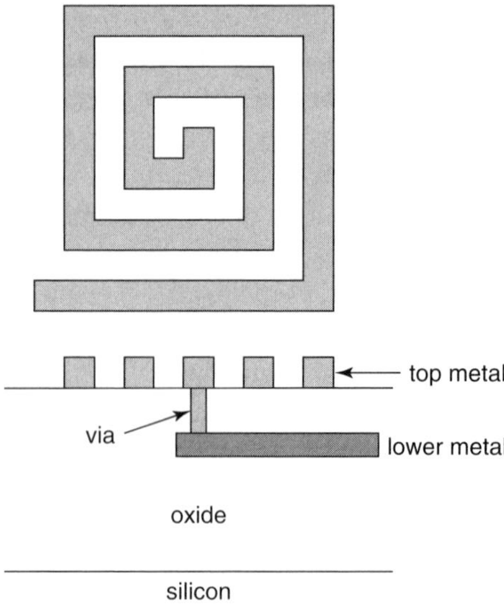

Figure 10.11 Plan and cross-section views of an integrated circuit inductor

currents that result from the inductor electric field cause power losses in the inductor and a lowering of the Q. Parasitic currents that result from the inductor magnetic field (eddy currents and image currents) not only cause power losses but also reduce the net inductance because of a reduction in the net magnetic field. Eddy currents and image currents in the substrate generally become significant for substrates with resistivities of less than 5 Ωcm [31]. Raising the substrate resistivity is one method of reducing substrate effects and hence of improving the inductor Q. This approach mimics that used in GaAs processes, where a semi-insulating substrate is available. This disadvantage of high-resistivity substrates is that significant changes to the CMOS process flow are needed to maintain the high resistivity. An alternative approach, which allows low-resistivity substrates to be used without inducing eddy and image currents, is to include a Faraday shield. This can take the form of a patterned ground plane [32] or a low-resistivity halo implant outside the inductor that forms a broken loop around the inductor [33]. Both of these approaches aim to terminate the parasitic electric field from the inductor in a low-impedance AC ground before it is able to penetrate any great distance into the substrate.

Varactor diodes are basically non-linear capacitors, and are used in circuits to provide frequency multiplication. Varactor diodes provide a non-linear capacitor by utilizing the non-linear capacitance/voltage characteristic of a junction diode. The main parameter of interest for a varactor diode is the maximum change in capacitance obtainable for a given change in voltage. A high value of this parameter can be obtained if an abrupt *pn* junction is used, with typical values in a SiGe BiCMOS technology being a factor of 1.7–1.8 for a 2.5 V change in voltage [34]. A heavily doped arsenic layer is generally used to produce a varactor diode because a very steep profile can be obtained due to concentration-enhanced diffusion. Faster diffusion is obtained at high arsenic concentrations than at low concentrations, which tends to square up the profile and give a large drop in arsenic concentration over a short distance.

For analogue circuit design, it is useful to have available *pnp* bipolar transistors so that complementary design approaches can be used. *pnp* bipolar transistors can be produced in a BiCMOS technology without any additional processing steps by placing two extrinsic base regions in close proximity and relying on the lateral injection of carriers to provide current gain. This arrangement is referred to as a lateral *pnp* transistor, and is illustrated in Figure 10.12. It is advantageous if the collector completely surrounds the emitter, since holes injected from the emitter can be collected on four sides, thereby maximizing the current gain. Additional improvements in gain can also be obtained if a buried layer is incorporated below the emitter of the *pnp* transistor. Holes injected vertically downwards from the emitter see a potential barrier at the *nn*+ junction and are reflected upwards. They can then diffuse to the collector and contribute to the collector current. Common emitter current gains

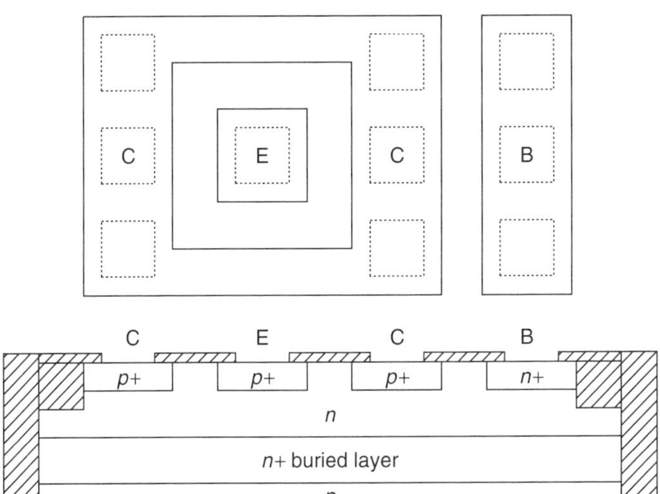

Figure 10.12 Plan and cross-section views of a *pnp* lateral bipolar transistor

of about 30 can be achieved in this way. The main disadvantage of lateral *pnp* transistors is a poor high-frequency performance, caused by the large amount of stored charge in the epitaxial base. If high-frequency *pnp* bipolar transistors are required, a complementary bipolar approach can be used, as discussed in Section 9.8 [35,36].

REFERENCES

[1] Special issue on heterostructure transistors, *IEEE Trans. Electron. Devices*, **36**, No. 10 (1989).

[2] Special issue on bipolar transistor technology; past and future trends, *IEEE Trans. Electron. Devices*, **48**, No. 11 (2001).

[3] J.-S. Rieh *et al.*, 'SiGe HBTs with cut-off frequency of 350 GHz', *IEDM Technical Digest*, **771** (2002).

[4] D.L. Harame, J.H. Comfort, J.D. Cressler, E.F. Crabbé, J.Y.-C. Sun, B.S. Meyerson and T. Tice, 'Si/SiGe epitaxial base transistors- part I: materials, physics and circuits', *IEEE Trans. Electron. Devices*, **42**, 455 (1995).

[5] D.L. Harame, J.H. Comfort, J.D. Cressler, E.F. Crabbé, J.Y.-C. Sun, B.S. Meyerson and T. Tice, 'Si/SiGe epitaxial base transistors- part II: process integration and analog applications', *IEEE Trans. Electron. Devices*, **42**, 469 (1995).

[6] F. Sato, T. Hashimoto, T. Tatsumi and T. Tashiro, 'Sub-20 ps ECL circuits with high performance super self-aligned selectively grown SiGe base bipolar transistors', *IEEE Trans. Electron. Devices*, **42**, 483 (1995).

[7] T.F. Meister, H. Schäfer, M. Franosch, W. Molzer, K. Aufinger, U. Scheler, C. Walz, S. Stolz, S. Boguth and J. Böck, 'SiGe base bipolar technology with 74 GHz f_{max} and 11 ps gate delay', *IEDM Technical Digest*, 739 (1995).

[8] A. Monroy, M. Laurens, M. Marty, D. Dutartre, D. Gloria, J.L. Carbonero, A. Perrotin, M. Roche and A. Chatre, 'BiCMOS6G: a high performance 0.35 μm SiGe BiCMOS technology for wireless applications', *Proc. BCTM*, 121 (1999).

[9] F. Sato, T. Hashimoto, T. Tatsumi and T. Tashiro, 'Sub-20 ps ECL circuits with high-performance super self-aligned selectively grown SiGe base bipolar transistors', *IEEE Trans. Electron. Devices*, **42**, 483 (1995).

[10] T.F. Meister, H. Schäfer, M. Franosch, W. Molzer, K. Aufinger, U. Scheler, C. Walz, M. Stolz, S. Boguth and J. Böck, 'SiGe base bipolar technology with 74 GHz f_{max} and 11 ps gate delay', *IEDM Technical Digest*, 739 (1995).

[11] K. Washio, E. Ohue, K. Oda, M. Tanabe, H. Shimamoto, T. Onai and M. Kondo, 'A selective epitaxial growth SiGe base HBT with SMI electrodes featuring 9.3 ps ECL gate delay', *IEEE Trans. Electron. Devices*, **46**, 1411 (1999).

[12] K. Washio, H. Shimamoto, K. Oda, R. Hayami, Y. Kiyota, M. Tanabe, M. Kondo, T. Hashimoto and T. Harade, 'A 0.2 μm 180GHz f_{max} 6.7 ps ECL SOI/HRS self-aligned SEG SiGe HBT/CMOS technology for microwave and high-speed digital applications', *IEEE Trans. Electron. Devices*, **49**, 271 (2002).

[13] A. Ishitani, H. Kitajima, N. Endo and N. Kasai, 'Facet formation in selective silicon epitaxial growth', *Japan Jnl Appl. Phys.*, **24**, 1267 (1985).

[14] C.I. Drowley and M.L. Hammond, 'Conditions for uniform selective epitaxial growth', *Solid State Technology*, **33**, 135 (1990).

[15] L.D. Lanzerotti, J.C. Sturm, E. Stach, R. Hull, T. Buyuklimanli and C. Magee, 'Suppression of boron transient enhanced diffusion in SiGe heterojunction bipolar transistors by carbon incorporation', *Appl. Phys. Lett.*, **23**, 3125 (1997).

[16] H.J. Osten, G. Lippert, D. Knoll, R. Barth, B. Heinemann, H. Rücker and P. Schley, 'The effect of carbon incorporation on SiGe heterojunction bipolar transistor performance and process margin', *IEDM Technical Digest*, 803 (1997).

[17] A. Gruhle, H. Kibbel and U. König, 'The reduction of base dopant out-diffusion in SiGe heterojunction bipolar transistors by carbon doping', *Appl. Phys. Lett.*, **75**, 1311 (1999).

[18] H. Rücker, B. Heinemann, Röpke, R. Kurps, D. Krüger, G. Lippert and H.J.Osten, 'Suppressed diffusion of boron and carbon in carbon-rich silicon', *Appl. Phys. Lett.*, **73**, 1682 (1998).

[19] B. Heinemann, D. Knoll, R. Barth, D. Bolze, K. Blum, J. Drews, K.-E. Ehwald, G.G. Fischer, K. Köpke, D. Krüger, R. Kurps, H. Rücker, P. Schley, W. Winkler and H.-E. Wulf, 'Cost effective high performance high-voltage SiGe:C HBTs with 100 GHz f_T and $BV_{CEO}f_T$ products exceeding 220VGHz', *IEDM Technical Digest* (2001).

[20] B. Martinet, H. Baudry, O. Kermarrec, Y. Campidelli, M. Laurens, M. Marty, T. Schwartzmann, A. Monroy, D. Bensahel and A. Chantre, '100 GHz SiGe:C HBTs using non selective base epitaxy', *Proc. ESSDERC*, 97 (2001).

[21] B. Jagannathan *et al.*, 'Self-aligned SiGe npn transistors with 285 GHz f_{max} and 207 GHz f_T in a manufacturable technology', *IEEE Electron. Device Lett.* **23**, 258 (2002).

[22] J. Böck, H. Schäfer, H. Knapp, D. Zöschg, K. Aufinger, M. Wurzer, S. Boguth, R. Stengl, R. Schreiter and T.F. Meister, 'High speed SiGe:C bipolar technology', *IEDM Technical Digest* (2001).

[23] K. Oda, E. Ohue, I. Suzumura, R. Hayami, A. Kodama, H. Shimamoto and K. Washio, 'Self-aligned selective epitaxial growth SiGeC HBT technology featuring 170 GHz f_{max}', *IEDM Technical Digest* (2001).

[24] S. Lombardo, A. Pinto, V. Raineri, P. Ward, G. La Rosa, G. Privitera and S.U. Campisano, 'Si/GeSi heterojunction bipolar transistors with GeSi base formed by Ge ion implantation in Si', *IEEE Electron. Device Letters*, **17**, 1 (1996).

[25] F. Cristiano, A. Nejim, D.A.O. Hope, M.R. Houlton and P.L.F. Hemment, 'Structural studies of ion beam synthesised SiGe/Si heterostructures for HBT applications', *Nucl. Inst. Methods Phys. Res. B*, **112**, 311 (1996).

[26] S. Lombardo, G. Privitera, A. Pinto, P. Ward, G. La Rosa and S.U. Campisano, 'Bandgap narrowing and high frequency characteristics of Si/GeSi heterojunction bipolar transistors formed by Ge ion implantation in Si', *IEEE Trans. Electron. Devices*, **45**, 1531 (1998).

[27] M.J. Mitchell, P. Ashburn, H. Graoui, P.L.F. Hemment, A. Lamb, S. Hall and S. Nigrin, 'A comparison on pnp and npn SiGe HBTs fabricated by Ge+ implantation', *Proc. ESSDERC*, 248 (2000).

[28] P.J. Ward; private communication.

[29] P.L.F. Hemment, F. Cristiano, A. Nejim, S. Lombardo, K.K. Larssen, F. Priolo and R.C. Barklie, 'Ge+ ion implantation: a competing technology', *Jnl Crystal Growth*, **157**, 147 (1995).

[30] J.N. Burghartz, M. Soyuer, K.A. Jenkins, M. Kies, M. Dolan, K.J. Stein, J. Malinowski and D.L. Harame, 'Integrated rf components in a SiGe BiCMOS technology', *IEEE Jnl Solid State Circuits*, **32**, 1440 (1997).

[31] D.L. Harame, D.C. Ahlgren, D.D. Coolbaugh, J.S. Dunn, G.G. Freeman, J.D. Gillis, R.A. Groves, G.N. Hendersen, R.A. Johnson, A.J. Joseph, S. Subbanna, A.M. Victor, K.M. Watson, C.S. Webster and P.J. Zampardi, 'Current status and future trends of SiGe BiCMOS technology', *IEEE Trans. Electron. Devices*, **48**, 2575 (2001).

[32] C.P. Yu and S.S. Wong, 'On-chip spiral inductors with patterned ground shields for Si-based RF ICs', *IEEE Jnl Solid State Circuits*, **33**, 743 (1998).

[33] J.N. Burghartz, A.E. Ruehli, K.A. Jenkins, M. Soyuer and D. Nguyen-Ngoc, 'Novel substrate contact structure for high Q silicon integrated spiral inductors', *IEDM Technical Digest*, 55 (1997).

[34] S.A. St. Onge *et al.*, 'A 0.24 μm SiGe BiCMOS mixed signal RF production technology featuring a 47 GHz f_T HBT and 0.18 μm L_{eff} CMOS', *Proc. IEEE BCTM*, 117 (1999).

[35] Y. Yoshida, H. Suzuki, Y. Kinoshita, H. Fujii and T. Yamazaki, 'An RF BiCMOS process using high f_{SR} spiral inductor with premetal deep trenches and a dual recessed bipolar collector sink', *IEDM Technical Digest*, 213 (1998).

[36] R. Bashir *et al.*, 'A complementary bipolar technology family with a vertically integrated pnp for high frequency analog applications', *IEEE Trans. Electron. Devices*, **48**, 2525 (2001).

11

Compact Models of Bipolar Transistors

11.1 INTRODUCTION

The efficient design of integrated circuits requires the use of sophisticated computer-aided circuit design programs such as the ubiquitous SPICE program. These programs take a circuit-level description as input and provide output in the form of node voltages and currents as a function of time. A vital component of a circuit simulation program is a compact transistor model, which defines the terminal characteristics of the transistor. Such a model consists of a combination of circuit elements such as resistors, capacitors, current generators, etc. and equations for defining the behaviour of the transistor. Definition of the transistor model is through a set of parameters, typically 40 for a full description of a bipolar transistor.

In devising a compact transistor model, an accurate description of the terminal characteristics is more important than a rigorous description of the device physics. Nevertheless, models that are based on the physics of the device do provide a better understanding, and can generally be implemented using fewer model parameters. For these reasons, most circuit simulators use compact transistor models that are to a first order physics-based, although second-order effects are often described using simple empirical expressions. Computational time is of paramount importance, since this provides a limit to the size of circuit that can be simulated. The need for short simulation times is the primary reason

SiGe Heterojunction Bipolar Transistors Peter Ashburn
 ISBN: 0-470-84838-3

that compact models are used for circuit simulation in preference to full numerical device simulation.

Compact transistor models provide an interface between process engineers, device engineers and integrated circuit designers. Circuit designers need to be familiar with compact models, because the accuracy of their circuit simulations depend critically on the accuracy of the transistor models and the associated input parameters. Similarly, process and device engineers need to have some knowledge of transistor models, because the transistor and process design need to be optimized to give optimum circuit performance. The relationship between process design, transistor design and circuit performance will be described in detail in Chapter 12.

In this chapter we will consider the compact bipolar transistor models that are used in widely available computer-aided circuit design programs such as SPICE [1]. The simple DC Ebers-Moll model [2] will be used as a starting point, and additional physical mechanisms added to the basic model as required. In this way, the full Gummel-Poon bipolar transistor model [3] will evolve in a number of well-defined and easy to understand stages. The SPICE bipolar transistor model will be described in detail, and the key features of more recent variants, such as the VBIC [4] and Mextram [5] models, will be briefly outlined.

11.2 EBERS-MOLL MODEL

The Ebers-Moll model [2] is a simple, large-signal model for describing the behaviour of a bipolar transistor. The DC model configuration is illustrated in Figure 11.1 for an *npn* transistor. Equations for the forward and reverse diode currents I_F and I_R are needed to complete the model, and these are:

$$I_F = I_{ES}\left(\exp\frac{qV_{BE}}{kT} - 1\right) \tag{11.1}$$

$$I_R = I_{CS}\left(\exp\frac{qV_{BC}}{kT} - 1\right) \tag{11.2}$$

A third equation, termed the reciprocity relation, links the saturation currents I_{ES} and I_{CS} to the common base current gains [6]:

$$\alpha_F I_{ES} = \alpha_R I_{CS} = I_S \tag{11.3}$$

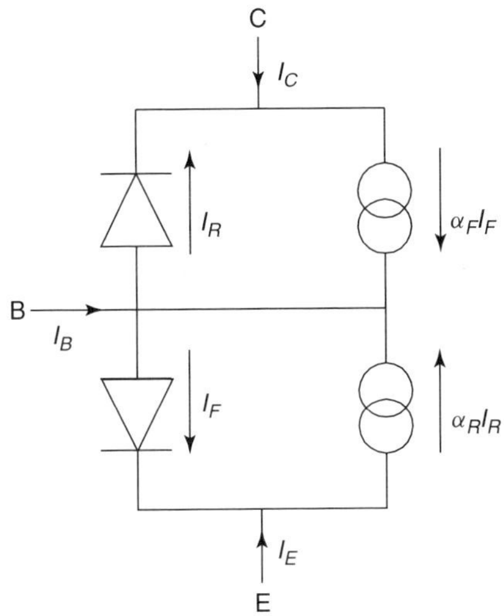

Figure 11.1 DC Ebers-Moll model of a bipolar transistor

From these equations it is clear that three parameters are needed to fully describe a transistor in the Ebers-Moll model, namely α_F, α_R and I_S. The terminal currents of the transistor can be easily expressed in terms of the three transistor parameters:

$$I_C = \alpha_F I_F - I_R \tag{11.4}$$

$$I_E = \alpha_R I_R - I_F \tag{11.5}$$

$$I_B = (1 - \alpha_F)I_F + (1 - \alpha_R)I_R \tag{11.6}$$

The Ebers-Moll model is firmly based on the physics of the device operation. The ideal diodes provide the expected exponential relationship between current and base/emitter voltage, and the current generators describe the transistor action. The compact model equations (11.1)–(11.3) are of the same form as the physically derived equations presented in Chapter 2.

All the components in Figure 11.1 are required to model a transistor in saturation, but in the forward and reverse active regions considerable simplifications can be made. In the forward active region, the collector/base diode is reverse biased, and hence both the collector/base diode and its associated current source can be omitted. Similarly in the reverse active region, the emitter/base diode is reverse biased and hence both the emitter/base diode and its associated current source can be omitted. The

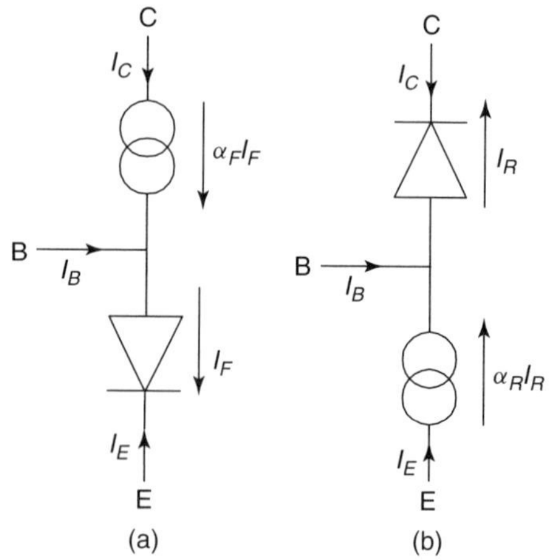

Figure 11.2 DC Ebers-Moll model in (a) forward active region and (b) reverse active region

models for these two situations are illustrated in Figure 11.2. Physically, in the forward active region the current I_F is the total current crossing the emitter/base junction ($(I_{ne} + I_{pe})$ in Figure 2.1), while $\alpha_F I_F$ is the electron current at the edge of the collector/base depletion region (I_{nc}). Recombination in the emitter/base depletion region (I_{rg}) is not modelled in the basic Ebers-Moll model.

11.3 NON-LINEAR HYBRID-π MODEL

For the purposes of computer simulation a change in the form of the Ebers-Moll model is desirable. The reason for this change is that it is difficult to model the base current accurately using the Ebers-Moll model because the emitter and collector currents are specified in the model and the base current is calculated from the difference between these two currents. The non-linear hybrid-π model is able to accurately model the base current because extra nonideal diodes can be added to model the effect of recombination in the emitter/base depletion region on the base current. The base and collector currents are explicitly specified in the non-linear hybrid-π model and the emitter current is calculated from these two currents.

In the non-linear hybrid-π model, the common component of current flowing from the emitter to the collector is identified, which allows a current generator I_{CT} to be specified as shown in Figure 11.3. From the Ebers-Moll model in Figure 11.2, this current can be identified as $\alpha_F I_F$ in

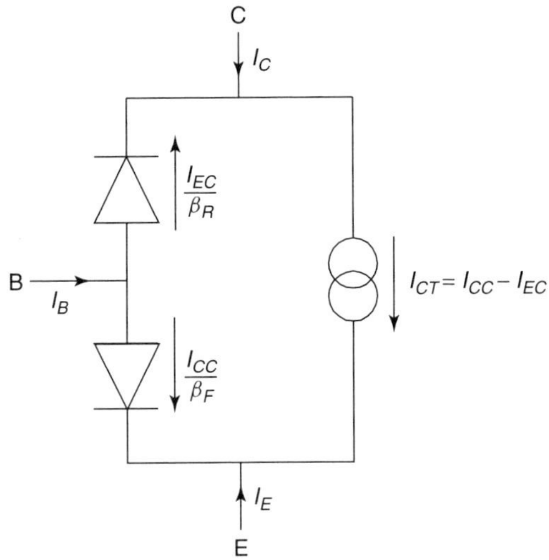

Figure 11.3 Non-linear hybrid-π model of a bipolar transistor

the forward active region and $-\alpha_R I_R$ in the reverse active region. When both junctions are forward biased, these two currents can be summed:

$$I_{CT} = I_{CC} - I_{EC} = \alpha_F I_F - \alpha_R I_R \tag{11.7}$$

$$= \alpha_F I_{ES}\left(\exp\frac{qV_{BE}}{kT} - 1\right) - \alpha_R I_{CS}\left(\exp\frac{qV_{BC}}{kT} - 1\right) \tag{11.8}$$

$$= I_S\left[\left(\exp\frac{qV_{BE}}{kT} - 1\right) - \left(\exp\frac{qV_{BC}}{kT} - 1\right)\right] \tag{11.9}$$

where the reciprocity relation has been used to obtain equation (11.9) and I_{CC} and I_{EC} are given by:

$$I_{CC} = I_S\left(\exp\frac{qV_{BE}}{kT} - 1\right) \tag{11.10}$$

$$I_{EC} = I_S\left(\exp\frac{qV_{BC}}{kT} - 1\right) \tag{11.11}$$

Using the Ebers-Moll equations (11.3)–(11.6), the equations for the terminal currents can be derived in terms of I_{CT}, β_F, β_R and I_S as:

$$I_C = I_{CT} - \frac{I_S}{\beta_R}\left(\exp\frac{qV_{BC}}{kT} - 1\right) \tag{11.12}$$

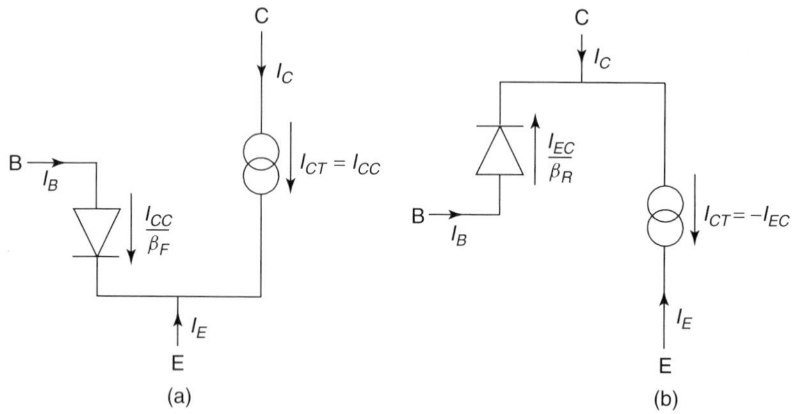

Figure 11.4 Non-linear hybrid-π model in (a) forward active region and (b) reverse active region

$$I_E = -I_{CT} - \frac{I_S}{\beta_F}\left(\exp\frac{qV_{BE}}{kT} - 1\right) \tag{11.13}$$

$$I_B = \frac{I_S}{\beta_F}\left(\exp\frac{qV_{BE}}{kT} - 1\right) + \frac{I_S}{\beta_R}\left(\exp\frac{qV_{BC}}{kT} - 1\right) \tag{11.14}$$

It can be clearly seen that these equations for the collector, emitter and base currents are the same as those given by the non-linear hybrid-π model in Figure 11.3. This serves to emphasize that the non-linear hybrid-π model is merely a rearrangement of the form of the Ebers-Moll model. Three parameters are needed to characterize a bipolar transistor in the non-linear hybrid-π model, and these are β_F, β_R and I_S.

In the forward and reverse active regions the model can be considerably simplified, as illustrated in Figure 11.4. In the forward active region, the collector/base diode is reverse biased, and hence the collector/base diode can be omitted and I_{CT} is approximately equal to I_{CC}. Similarly in the reverse active region, the emitter/base diode is reverse biased and hence the emitter/base diode can be omitted and I_{CT} is approximately equal to $-I_{EC}$.

11.4 MODELLING THE LOW-CURRENT GAIN

As discussed in Section 4.2, recombination of minority carriers in the emitter/base depletion region gives rise to a nonideal, $\exp(qV_{BE}/(mKT))$ dependence of the base current. This behaviour can be easily modelled in

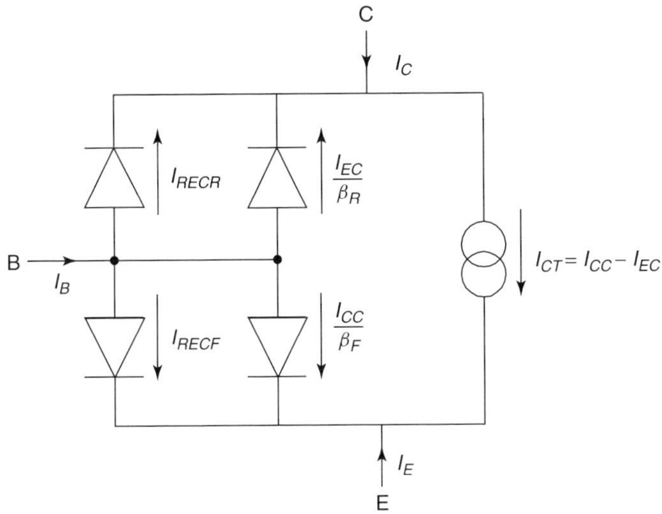

Figure 11.5 Modelling of the low-current gain in the non-linear hybrid-π model

the non-linear hybrid-π model by adding extra nonideal diodes I_{RECF} and I_{RECR} in parallel with the ideal diodes, as shown in Figure 11.5. The first diode I_{RECF} models recombination in the forward biased emitter/base depletion region in forward active operation and the second diode I_{RECR} models recombination in the forward biased collector/base depletion region in reverse active operation.

The equations for I_{RECF} and I_{RECR} take the following form:

$$I_{RECF} = I_{SE}\left(\exp\frac{qV_{BE}}{N_E kT} - 1\right) \tag{11.15}$$

$$I_{RECR} = I_{SC}\left(\exp\frac{qV_{BC}}{N_C kT} - 1\right) \tag{11.16}$$

where N_E and I_{SE} are the emitter/base recombination ideality factor and saturation current, respectively, in forward active operation, and N_C and I_{SC} are the equivalent in reverse active operation. Four model parameters are needed to completely specify the low-current gain, and these are I_{SE}, N_E, I_{SC} and N_C. These parameters can be measured from a Gummel plot, as illustrated in Figure 11.6 for the case of forward active operation. Also shown is the method of measuring the parameters β_F and I_S. A Gummel plot measurement for reverse operation will yield an equivalent set of parameters for reverse operation.

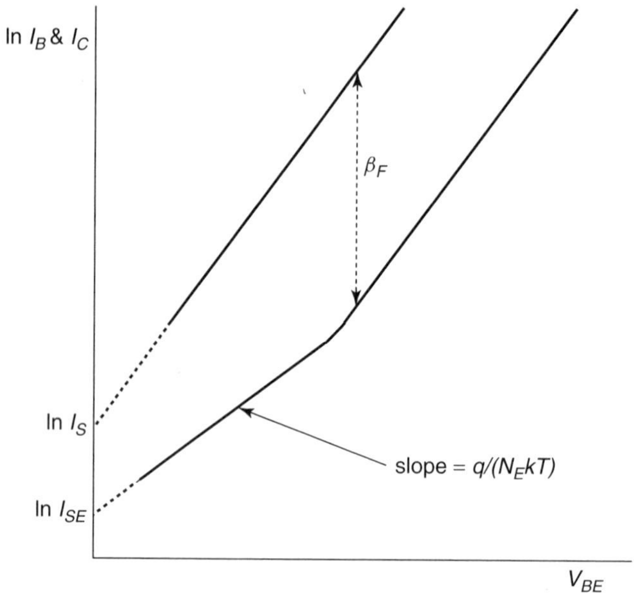

Figure 11.6 Measurement of the compact model parameters I_S, β_F, I_{SE}, and N_E from a Gummel plot measured under forward active operation

11.5 AC NON-LINEAR HYBRID-π MODEL

The bipolar transistor models discussed so far are only suitable for modelling the DC characteristics of bipolar transistors. In order to model AC effects, charge storage in the device must be described, and this requires the incorporation of additional parameters. These can be characterized into three broad types: series ohmic resistances, depletion capacitances and charge storage capacitances due to the mobile carriers in the transistor (diffusion capacitances). Figure 11.7 shows how the DC model can be extended to include these additional AC parameters. Note that internal nodes E', B' and C' have been defined and hence the equations discussed above in Sections 11.3 and 11.4 now have to be written in terms of these internal node voltages.

The resistors R_C, R_E and R_B represent the series resistance of the semiconductor between the active transistor area and the emitter, collector and base contacts, as discussed in Section 5.6, and the capacitors C_{JC} and C_{JE} represent the collector/base and emitter/base depletion capacitances, as discussed in Section 5.7. The capacitors labelled Q_{DC} and Q_{DE} represent the charge due to the mobile carriers in the transistor. This charge

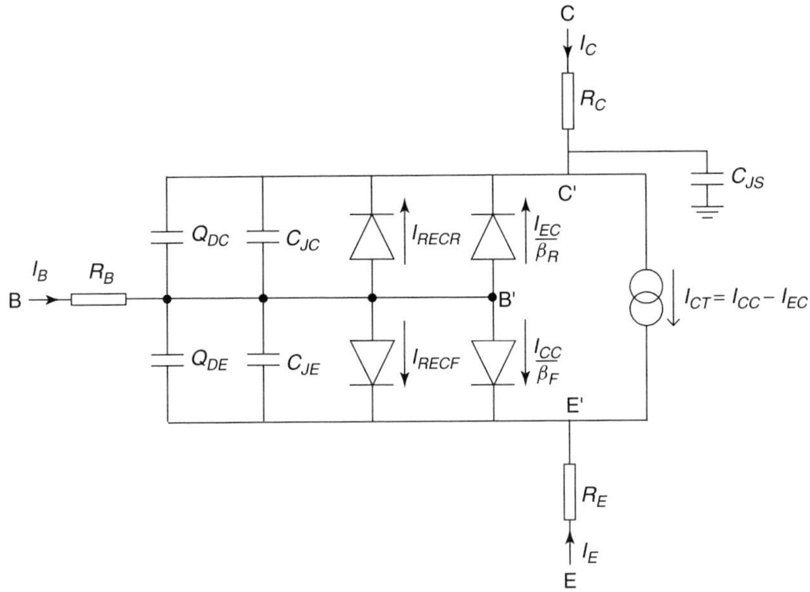

Figure 11.7 The AC non-linear hybrid-π model of a bipolar transistor

can be conveniently partitioned into two parts, one associated with the forward-biased emitter/base junction and one with the forward- biased collector/base junction. In forward active operation, the stored charge associated with the forward biased emitter/base junction Q_{DE} can be related to the forward transit time τ_F:

$$Q_{DE} = \tau_F I_{CC} \tag{11.17}$$

In reverse active operation the stored charge associated with the forward biased collector/base junction Q_{DC} can similarly be related to the reverse transit time τ_R:

$$Q_{DC} = \tau_R I_{EC} \tag{11.18}$$

For a transistor in saturation, both emitter and collector junctions are forward biased. The total minority carrier charge in the transistor can therefore be calculated by assuming that superposition applies. In other words, the total minority carrier charge is assumed to be equal to the sum of the charge due to each junction acting separately, i.e. $Q_{DE} + Q_{DC}$. These charges can be related to the non-linear diffusion capacitances C_{DE} and C_{DC} by:

$$C_{DE} = \frac{Q_{DE}}{V_{B'E'}} \tag{11.19}$$

$$C_{DC} = \frac{Q_{DC}}{V_{B'C'}} \tag{11.20}$$

Two parameters are needed to model the stored charge, namely the forward and reverse transit times τ_F and τ_R. These parameters can be measured from an f_T measurement, as described in Section 5.3.

11.6 SMALL-SIGNAL HYBRID-π MODEL

The model in Figure 11.7 is non-linear, and hence circuit analysis can only proceed with the aid of a computer. However, in circuits where the AC signal excursions around the quiescent operating point are small, it is possible to approximate the non-linear elements in Figure 11.7 by linear elements. In this situation, a small-signal model is obtained that can be applied to a variety of analogue circuits, amplifiers being one example.

For the majority of small-signal applications, the model in Figure 11.7 can be considerably simplified and linearized to produce a small-signal model. In the forward active region the emitter/base junction is forward biased and the collector/base reverse biased. There is therefore no charge storage associated with the collector junction, and hence the collector/base diode and its associated diffusion capacitance can be omitted. The series resistances R_C, R_E and R_B, the substrate capacitance C_{JS} and recombination in the depletion region can also to a first order be neglected. These approximations lead to the simplified version of the non-linear AC hybrid-π model in Figure 11.8(a).

The forward-biased emitter/base diode in Figure 11.8(a) can be linearized to produce an equivalent input resistance r_π that can be derived by differentiating the base current with respect to the base/emitter voltage:

$$I_B \approx \frac{I_S}{\beta_F} \exp \frac{qV_{BE}}{kT} \tag{11.21}$$

$$\frac{\partial I_B}{\partial V_{BE}} = \frac{I_S}{\beta_F}\frac{q}{kT} \exp \frac{qV_{BE}}{kT} = \frac{qI_B}{kT} \tag{11.22}$$

$$r_\pi = \frac{\partial V_{BE}}{\partial I_B} = \frac{kT}{qI_B} \tag{11.23}$$

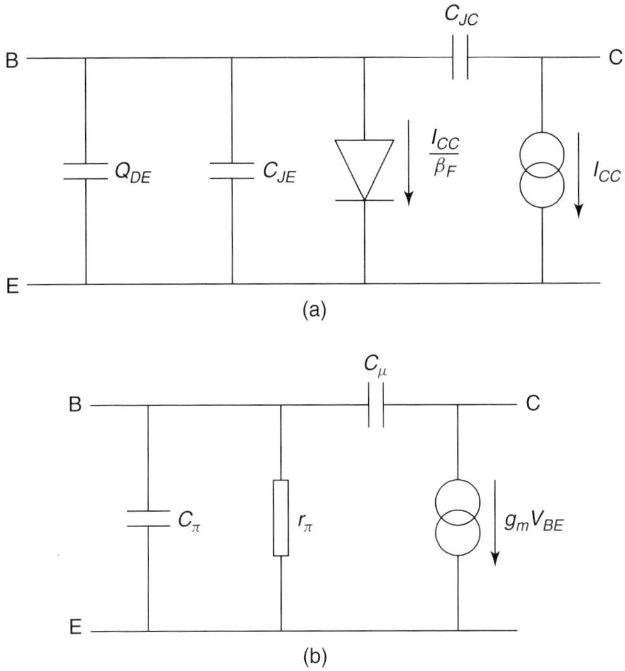

Figure 11.8 Simplified version of hybrid-π model; (a) non-linear hybrid-π model; (b) small-signal hybrid-π model

Similarly, the current generator in Figure 11.8(a) can be linearized by differentiating the collector current with respect to the base/emitter voltage:

$$I_C \approx I_S \exp \frac{qV_{BE}}{kT} \tag{11.24}$$

$$\frac{\partial I_C}{\partial V_{BE}} = g_m = I_S \frac{q}{kT} \exp \frac{qV_{BE}}{kT} = \frac{qI_C}{kT} \tag{11.25}$$

where g_m is the transconductance of the transistor. Finally, the emitter diffusion capacitance C_{DE} can be linearized to:

$$Q_{DE} = \tau_F I_C \tag{11.26}$$

$$C_{DE}(\text{small signal}) = \frac{\partial Q_{DE}}{\partial V_{BE}} = \tau_F \frac{q}{kT} I_S \exp \frac{qV_{BE}}{kT} = g_m \tau_F \tag{11.27}$$

where equation (11.17) has been used for Q_{DE}. The resulting small-signal equivalent circuit is shown in Figure 11.8(b). The capacitor C_μ

is the collector/base depletion capacitance and C_π is the sum of the emitter/base depletion capacitance and the emitter diffusion capacitance:

$$C_\pi = C_{JE} + g_m \tau_F \tag{11.28}$$

11.7 GUMMEL-POON MODEL

The Gummel-Poon model [3] was introduced in 1970 and is an improved version of the AC non-linear hybrid-π model in Figure 11.7. Two second-order, high-level effects are modelled in an elegant and unified way:

(1) high-level injection;

(2) basewidth modulation.

The Gummel-Poon model has been described in detail in the literature [3,6], and hence in this section we will merely state the relevant model equations, without considering their derivation. This will allow the emphasis of this section to be directed towards explanations of the underlying physical justification of the equations.

The essence of the Gummel-Poon model is a new definition of the current I_{CT} in terms of the internal physics of the transistor:

$$I_{CT} = \frac{I_S}{Q_B}\left[\left(\exp\frac{qV_{B'E'}}{kT} - 1\right) - \left(\exp\frac{qV_{B'C'}}{kT} - 1\right)\right] \tag{11.29}$$

where Q_B is the majority carrier charge in the base, normalized to the zero-bias majority carrier charge in the base. At zero bias Q_B is therefore equal to unity, and equation (11.29) reduces to equation (11.9). On application of bias to the junctions, Q_B takes on values other than unity. This provides a means of modelling basewidth modulation, high-level injection and the variation of τ_F with I_C, since all these mechanisms modulate the majority carrier charge in the base.

The normalized majority carrier charge in the base Q_B is given by an equation of the form:

$$Q_B = \frac{Q_1}{2} + \sqrt{\left(\frac{Q_1}{2}\right)^2 + Q_2} \tag{11.30}$$

where

$$Q_1 = \frac{1}{1 - \dfrac{V_{B'C'}}{V_{AF}} - \dfrac{V_{B'E'}}{V_{AR}}} \tag{11.31}$$

and

$$Q_2 = \frac{I_S}{I_{KF}}\left(\exp\frac{qV_{B'E'}}{kT} - 1\right) + \frac{I_S}{I_{KR}}\left(\exp\frac{qV_{B'C'}}{kT} - 1\right) \tag{11.32}$$

V_{AF} is the forward Early voltage as defined in Figure 4.12 and V_{AR} is an equivalent reverse Early voltage, which needs to be modelled when the emitter/base junction is reverse biased. I_{KF} is the forward knee current which defines the onset of high-level injection, and can be measured from a Gummel plot as illustrated in Figure 11.9. I_{KR} is an equivalent reverse knee current.

The physical significance of equations (11.30)–(11.32) can be understood by considering the simplified case of a device in the forward active region. In this case equations (11.29), (11.31) and (11.32) reduce to:

$$I_{CT} = \frac{I_S}{Q_B}\left(\exp\frac{qV_{B'E'}}{kT} - 1\right) \tag{11.33}$$

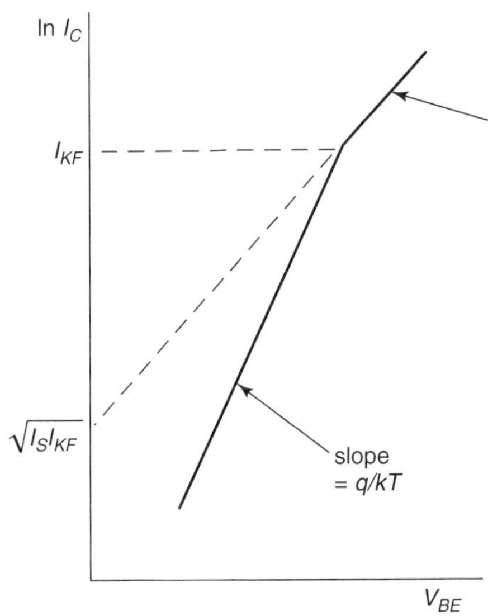

Figure 11.9 Gummel plot showing the knee current that defines the onset of high-level injection

$$Q_1 = \frac{1}{1 - \dfrac{V_{B'C'}}{V_{AF}}} \tag{11.34}$$

$$Q_2 = \frac{I_S}{I_{KF}} \left(\exp \frac{qV_{B'E'}}{kT} - 1 \right) \tag{11.35}$$

We will first consider the case of high-level injection. The criterion for the onset of high-level injection in the Gummel-Poon model is:

$$Q_2 \gg \frac{Q_1^2}{4} \tag{11.36}$$

Under high-level injection conditions the normalized majority carrier charge in the base Q_B can therefore be approximated by:

$$Q_B = \sqrt{Q_2} \tag{11.37}$$

Substituting equations (11.37) and (11.35) into equation (11.33) yields:

$$I_{CT} = \sqrt{I_S I_{KF}} \exp \frac{qV_{B'E'}}{2kT} \tag{11.38}$$

This equation gives the expected $\exp(qV_{BE}/(2kT))$ dependence of the collector current in the high-level injection regime, as predicted by equation (4.24) in Chapter 4. The intercept with the current axis is $\sqrt{I_S I_{KF}}$ as shown in Figure 11.9.

When the device is operating in low-level injection $Q_2 \ll Q_1^2/4$ and hence $Q_B \approx Q_1$. Using this approximation and substituting equation (11.34) into equation (11.33) gives:

$$I_{CT} = I_S \left(1 - \frac{V_{B'C'}}{V_{AF}} \right) \left(\exp \frac{qV_{B'E'}}{kT} - 1 \right) \tag{11.39}$$

This equation has the required $\exp(qV_{BE}/(kT))$ dependence, but is multiplied by the term in parentheses which models basewidth modulation. The physical significance of this additional term is illustrated in Figure 11.10. The collector current at a given collector/emitter voltage is the sum of the collector current at zero collector/base volts $I_C(0)$ and

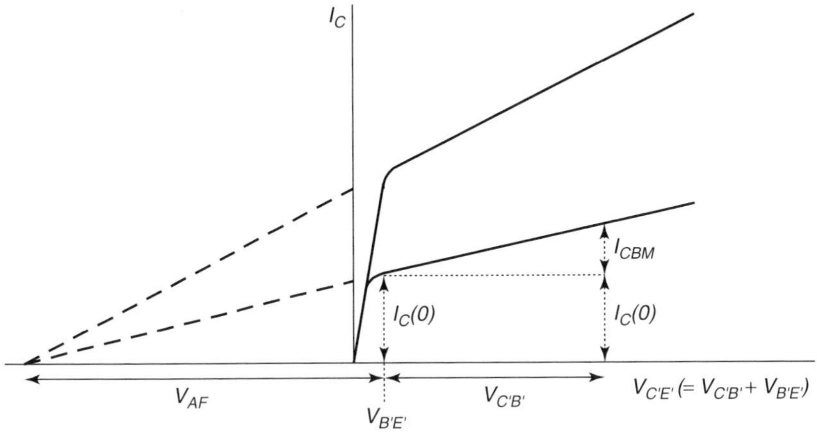

Figure 11.10 Transistor output characteristics illustrating the forward Early voltage V_{AF}

that due to basewidth modulation I_{CBM}. From Figure 11.10 it can be seen that the collector current is given by:

$$I_C = I_C(0) + I_{CBM} = I_C(0) + I_C(0)\frac{V_{C'B'}}{V_{AF}} \tag{11.40}$$

$$= I_C(0)\left(1 + \frac{V_{C'B'}}{V_{AF}}\right) \tag{11.41}$$

Equation (11.41) is exactly the same form as the Gummel-Poon model equation (11.39). The change in sign comes about because in one case the collector/base voltage is defined with respect to the base, while in the other it is defined with respect to the collector.

The Gummel-Poon model requires four additional parameters to model the Early voltage and high-level injection, namely V_{AF}, V_{AR}, I_{KF}, and I_{KR}. Also, in measuring the saturation current I_S the Gummel plot must be taken at a collector/base voltage of 0 V so that basewidth modulation is properly modelled. The four model parameters can easily be measured from Gummel plots and transistor output characteristics, as shown in Figures 11.9 and 11.10.

11.8 THE SPICE BIPOLAR TRANSISTOR MODEL

The SPICE circuit simulation program was introduced in 1973, and is now widely used throughout the world for the simulation of integrated circuits. It is a circuit simulator, which means that the device models need to be as simple as possible in order to minimize computational time, and hence allow relatively complex circuits to be simulated. The program has built-in device models for bipolar transistors, MOSFETS and JFETS, and input to these models is through sets of transistor parameters. The parameters for the SPICE bipolar transistor model are summarized in Table 11.1 and the equivalent circuit is shown in Figure 11.11.

The SPICE compact bipolar transistor model is essentially a Gummel-Poon model with a few minor modifications. Full details of the model are given below, together with a description of methods used to measure the parameter values.

11.8.1 Collector Current and Base Current

The full equations for the collector and base currents in the SPICE bipolar transistor model are as follows:

$$I_C = \frac{I_S}{Q_B}\left[\left(\exp\frac{qV_{B'E'}}{N_F kT} - 1\right) - \left(\exp\frac{qV_{B'C'}}{N_R kT} - 1\right)\right] - \frac{I_S}{\beta_R}\left(\exp\frac{qV_{B'C'}}{N_R kT} - 1\right) - I_{SC}\left(\exp\frac{qV_{B'C'}}{N_C kT} - 1\right) \quad (11.42)$$

$$I_B = \frac{I_S}{\beta_F}\left(\exp\frac{qV_{B'E'}}{N_F kT} - 1\right) + \frac{I_S}{\beta_R}\left(\exp\frac{qV_{B'C'}}{N_R kT} - 1\right) + I_{SE}\left(\exp\frac{qV_{B'E'}}{N_E kT} - 1\right) + I_{SC}\left(\exp\frac{qV_{B'C'}}{N_C kT} - 1\right) \quad (11.43)$$

where Q_B is given by equations (11.30)–(11.32). Two additional parameters N_F and N_R have been introduced to allow the exponents of the ideal emitter/base and collector/base diodes to be altered. In most practical Si bipolar transistors and SiGe HBTs these parameters would be set equal to unity. The parameters in equations (11.42) and (11.43) are measured from forward and reverse Gummel plots, as described in Sections 11.4 and 11.7, and from the transistor output characteristics, as described in Section 11.7.

11.8.2 Forward Transit Time

The forward transit time increases at high collector currents as a result of the Kirk effect, as discussed in Section 5.5 and as illustrated in

Table 11.1 Basic SPICE 2G bipolar transistor model parameters [1]

Basic DC parameters	
IS	Saturation current
BF	Maximum ideal forward gain
BR	Maximum ideal reverse gain
NF	Forward current ideality factor NR reverse current ideality factor
Basic AC parameters	
RC	Collector resistance
RE	Emitter resistance
RB	Low-current base resistance
IRB	Current where base resistance falls halfway to its maximum value
RBM	High-current base resistance
CJE0	Emitter/base, zero bias depletion capacitance
VJE	Emitter/base built-in voltage
MJE	Emitter/base profile exponent
CJC0	Base/collector, zero bias depletion capacitance
VJC	Base/collector built-in voltage
MJC	Base/collector profile exponent
XCJC	Fraction of B/C depletion capacitance connected to internal base node
CJS0	Collector/substrate, zero bias capacitance
VJS	Collector/substrate built-in voltage
MJS	Collector/substrate profile exponent
FC	Coefficient for depletion capacitances in forward bias
TF	Forward transit time
TR	Reverse transit time
Gummel-Poon parameters	
IKF	Knee current for roll-off of forward gain at high currents
IKR	Knee current for roll-off of reverse gain at high currents
VAF	Forward Early voltage
VAR	Reverse Early voltage
XTF	Coefficient for bias dependence of TF
VTF	Voltage describing V_{BC} dependence of TF
ITF	Parameter for variation of TF at high currents
ISE	Saturation current for base/emitter leakage current
NE	Low-current forward current ideality factor
ISC	Saturation current for base/collector leakage current
NC	Low-current reverse current ideality factor
Additional parameters	
EO	Semiconductor bandgap for temperature dependence of IS
XTI	Temperature exponent for effect on IS
XTB	Forward and reverse gain temperature exponent
PTF	Excess phase in g_m generator at frequency of $1/(2\pi\tau_F)$ Hz
KF	Flicker noise coefficient
AF	Flicker noise exponent

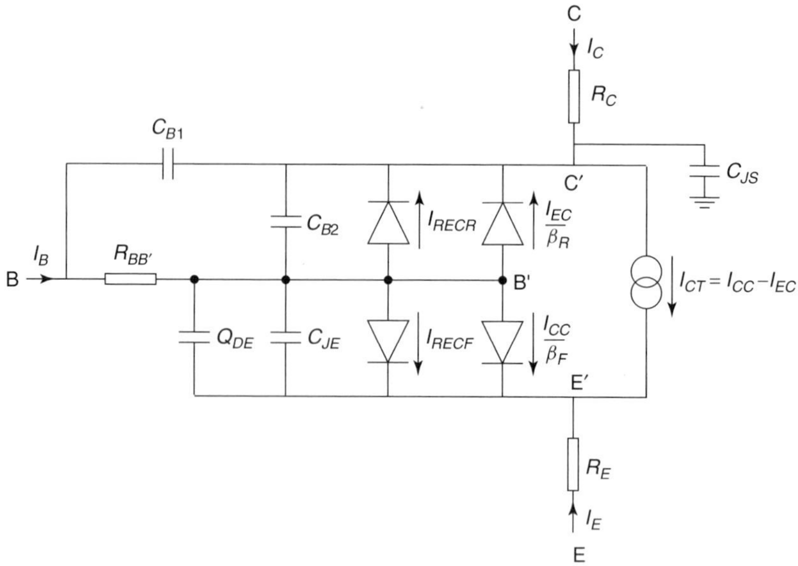

Figure 11.11 SPICE compact bipolar transistor model [1]

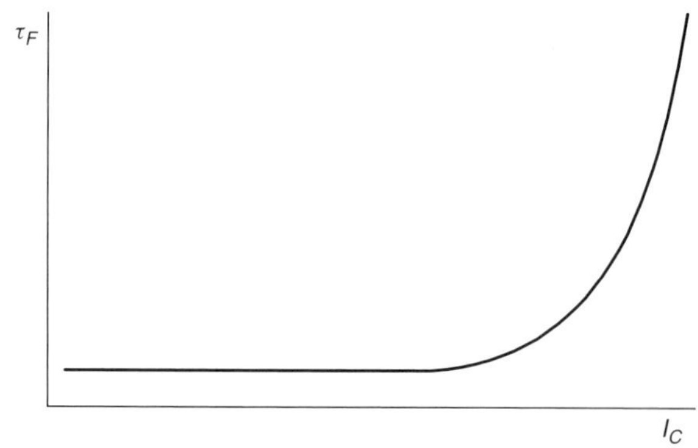

Figure 11.12 Variation of forward transit time with collector current due to the Kirk effect

Figure 11.12. In the SPICE 2G model, the Kirk effect is modelled differently than as originally proposed by Gummel and Poon [3]. The effect is modelled using an empirical equation to give greater flexibility in fitting the measured dependence of forward transit time τ_F on collector current:

$$\tau_F = T_F \left[1 + X_{TF} \left(\frac{I_{CC}}{I_{CC} + I_{TF}} \right)^2 \exp \frac{V_{B'C'}}{1.44\, V_{TF}} \right] \tag{11.44}$$

where T_F is the SPICE parameter that defines the value of forward transit time at low currents. Three parameters are used in this expression to model the behaviour of the forward transit time at high currents, namely X_{TF}, I_{TF} and V_{TF}. X_{TF} models the magnitude of the Kirk effect, I_{TF} the dependence on current and V_{TF} the dependence on collector/base voltage. This latter dependence occurs because the basewidth varies with collector/base voltage through the Early effect (Section 4.4), which in turn affects the base transit time and the forward transit time, as can be seen from equations (5.4) and (5.1).

The forward transit time τ_F is determined from a measurement of the cut-off frequency f_T, as described in Section 5.3. The parameters in equation (11.44) are determined by fitting to the measured dependence of forward transit time on collector current.

11.8.3 Base Resistance

The SPICE base resistance model includes current crowding, as discussed in Section 5.9, and conductivity modulation due to high-level injection, as discussed in Section 4.3. Both of these effects cause the base resistance $R_{BB'}$ to be current dependent. In the SPICE 2G model, the current dependence of the base resistance is modelled by the following simple equation:

$$R_{BB'} = R_{BM} + \frac{R_B - R_{BM}}{Q_B} \tag{11.45}$$

The current dependence of $R_{BB'}$ arises from the variation of Q_B with current. At high currents Q_B is very large, and equation (11.45) reduces to $R_{BB'} = R_{BM}$. The parameter R_{BM} therefore represents the high-current value of base resistance, which is essentially the extrinsic base resistance. At low currents, Q_B is equal to Q_1 and in the absence of basewidth modulation Q_1 is equal to unity. Equation (11.45) therefore reduces to $R_{BB'} = R_B$. The parameter R_B is therefore the low-current value of base resistance, namely, the sum of the intrinsic and extrinsic base resistances.

A variety of base resistance models have been developed to accurately model the variation of base resistance with current for different types of bipolar transistor. These models can often be accessed in different versions of SPICE. In the SPICE 2G model an alternative base resistance model is available, which can be chosen using the parameter I_{RB}. With $I_{RB} = 0$, equation (11.45) is used, whereas with $I_{RB} > 0$, the following equations are used:

$$R_{BB'} = R_{BM} + 3(R_B - R_{BM})\left(\frac{\tan(z) - z}{z\tan^2(z)}\right) \tag{11.46}$$

$$z = \frac{-1 + \sqrt{1 + \dfrac{144 I_B}{\pi^2 I_{RB}}}}{\dfrac{24}{\pi^2}\sqrt{\dfrac{I_B}{I_{RB}}}} \tag{11.47}$$

The parameter I_{RB} represents the current at which the base resistance falls to half of its minimum value. In general, equation (11.45) gives the best modelling of the base resistance when conductivity modulation dominates, whereas equation (11.46) gives the best modelling when current crowding dominates.

The base resistance of a bipolar transistor is an extremely important parameter because it has a strong influence on the high-frequency performance. A wide variety of methods have been devised for measuring the base resistance, and the value obtained depends not only on the measurement method, but also on the conditions used for the measurement [6]. The measurement methods can be broadly partitioned into three types, namely small-signal techniques [7], pulse techniques [8] and noise measurement techniques [9]. In general, it is advisable to choose a measurement technique that matches the circuit application. For the majority of applications, a small-signal measurement technique is the most appropriate and the input impedance circle method is the classical approach used [7]. This method assumes the small-signal hybrid-π model and measures the input impedance as a function of frequency, with the AC collector voltage kept at zero. When the impedance is plotted on the complex impedance plane, the locus of points should form a circle. The impedance value at the left intercept, which occurs at high frequencies, gives the value of $R_{BB'}$. This method is reasonably accurate at high values of collector current [7], but requires a lot of detailed measurements.

11.8.4 Collector Resistance

The collector resistance is modelled in SPICE 2G using a constant resistor, though in practice it varies with current because of conductivity modulation of the collector in heavy saturation. The collector resistance can be measured from the output characteristic of the bipolar transistor,

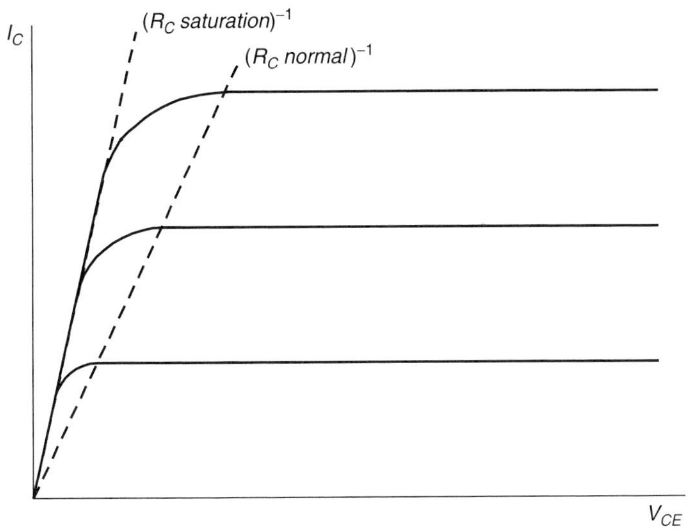

Figure 11.13 Measurement of collector resistance of a bipolar transistor

as illustrated in Figure 11.13. For transistors operating in the normal forward active region, the collector resistance can be obtained by drawing a straight line through the point of each curve where it deviates from a straight line. The reciprocal of the slope of this line then gives the collector resistance in the forward active region of operation (R_C normal). For transistors operating in strong saturation, a lower value of collector resistance is appropriate, as illustrated in Figure 11.13 (R_C saturation line).

11.8.5 Emitter Resistance

The emitter resistance in SPICE 2G is modelled using a constant resistor. It generally has a value of just a few ohms, but can be larger in small-geometry polysilicon emitters, as discussed in Section 6.7. The emitter resistance can be determined from a measurement of the base current as a function of collector/emitter voltage for a transistor with an open-circuit collector [10], as shown in Figure 11.14. The characteristic gives a straight line at low currents, but deviates from a straight line at high currents, and hence measurements should be taken at low currents. The emitter resistance can also be measured from the deviation of the Gummel plot from ideality at high currents [11], as discussed in Section 4.5.

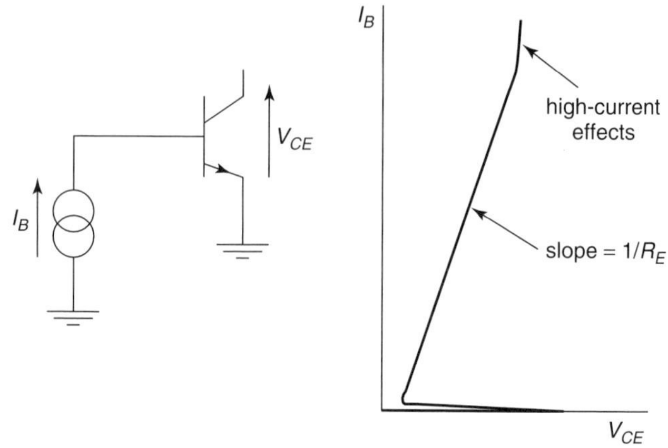

Figure 11.14 Measurement of emitter resistance using the open collector method

11.8.6 Emitter, Collector and Substrate Capacitances

The emitter/base depletion capacitance C_{JE} is modelled using the standard equation for the capacitance of a *pn* junction:

$$C_{JE} = \frac{C_{JE0}}{\left(1 - \frac{V_{B'E'}}{V_{JE}}\right)^{M_{JE}}} \tag{11.48}$$

where C_{JE0} is the emitter/base depletion capacitance at zero bias, V_{JE} is the emitter/base junction built-in voltage and M_{JE} is the emitter/base profile exponent that defines the sharpness of the emitter/base doping profile. This parameter has a value of 0.5 for an abrupt profile and 0.33 for a linearly graded profile. A similar expression is used for the collector/substrate depletion capacitance.

The base/collector depletion and diffusion capacitances have been combined in the SPICE 2G model into a total base/collector capacitance C_{BC} given by:

$$C_{B1} = C_{BC}(1 - X_{CJC}) \tag{11.49}$$

$$C_{B2} = C_{BC}X_{CJC} \tag{11.50}$$

$$C_{BC} = \frac{C_{JC0}}{\left(1 - \frac{V_{B'C'}}{V_{JC}}\right)^{M_{JC}}} + \frac{T_R I_{EC}}{V_{B'C'}} \tag{11.51}$$

The parameter X_{CJC} allows the distributed nature of the base resistance and base/collector capacitance to be modelled. When X_{CJC} is set equal to unity, C_{B2} becomes equal to the total base/collector capacitance C_{BC}. In this case, the modelling of the base/collector capacitance in the SPICE 2G model of Figure 11.11 is equivalent to that in Figure 11.7. Alternatively, the collector capacitance can be partitioned into an extrinsic and intrinsic component. In this case, X_{CJC} can be chosen to make C_{B2} equal to the intrinsic collector capacitance and C_{B1} equal to the extrinsic collector capacitance.

11.8.7 Additional Parameters

In the SPICE 2G model there are a number of additional parameters used to model noise and excess phase in the bipolar transistor, as illustrated in Table 11.1. The excess phase parameter, P_{TF}, models the extra phase shift that is obtained as a result of the time delay in the base and its distributed nature. The measured phase shift is larger than predicted by the poles of the equivalent circuit, and hence the parameter P_{TF} allows an extra phase shift to be added to the collector current to account for this effect. The parameters K_F and A_F are used to model $1/f$ noise in the bipolar transistor. Additional parameters are also used to model the temperature dependence of the transistor behaviour.

11.9 LIMITATIONS OF THE SPICE BIPOLAR TRANSISTOR MODEL

The Gummel-Poon model that is at the heart of the SPICE bipolar transistor model has been the most popular model for the design of bipolar circuits for a considerable period of time. The reason for its success is the wide range of mechanisms that are implemented in a simple and elegant way. However, since the development of the SiGe HBT, telecommunications circuits have been designed that operate at much higher frequencies and at lower supply voltages than has previously been the case. While SiGe HBTs can be modelled using the same basic approach as Si bipolar transistors, the high frequency and low voltage operation of many telecommunications circuits have uncovered a number of limitations in the Gummel-Poon model. These limitations mainly relate to the modelling of the collector epitaxial layer, and can be summarized as follows:

- quasi-saturation is not modelled;
- poor modelling of substrate effects;

- f_T roll-off at high currents (Kirk effect) modelled by curve fitting;
- no modelling of avalanche effects;
- early voltages are constant;
- poor modelling of temperature dependence of transistor behaviour.

Over the past few years, two new public domain [12,13] bipolar transistor models have been developed, namely the VBIC [4] and the Mextram models [5]. These new models include improved modelling of the epitaxial layer of the bipolar transistor based on the approach of Kull *et al.* [14]. Brief descriptions of the VBIC and Mextram models are given in the following sections. These models are being continually developed and the interested reader is referred to [12] and [13] for current summaries of the status of these models.

11.10 VBIC MODEL

The VBIC model is designed to be as similar to the Gummel-Poon model as possible, but includes the following improvements:

- modelling of quasi-saturation;
- parasitic substrate transistor modelled;
- avalanche multiplication modelled;
- improved Early effect modelling;
- improved temperature modelling;
- parasitic fixed (oxide) capacitance modelling;
- electrothermal modelling;
- base current is decoupled from collector current.

The equivalent circuit of the VBIC model is illustrated in Figure 11.15. Substrate effects are modelled using a substrate pnp bipolar transistor. Two sets of Gummel-Poon parameters are therefore required, one for the *npn* transistor and one for the parasitic *pnp* transistor, as shown by the dashed boxes in Figure 11.15. The variable resistor R_{BP} models the effects of current crowding and conductivity modulation on the intrinsic base resistance of the parasitic *pnp* transistor in the same way as the variable resistor R_{BI} models these effects in the *npn* transistor. The sidewall of the emitter/base junction in the *npn* transistor is modelled

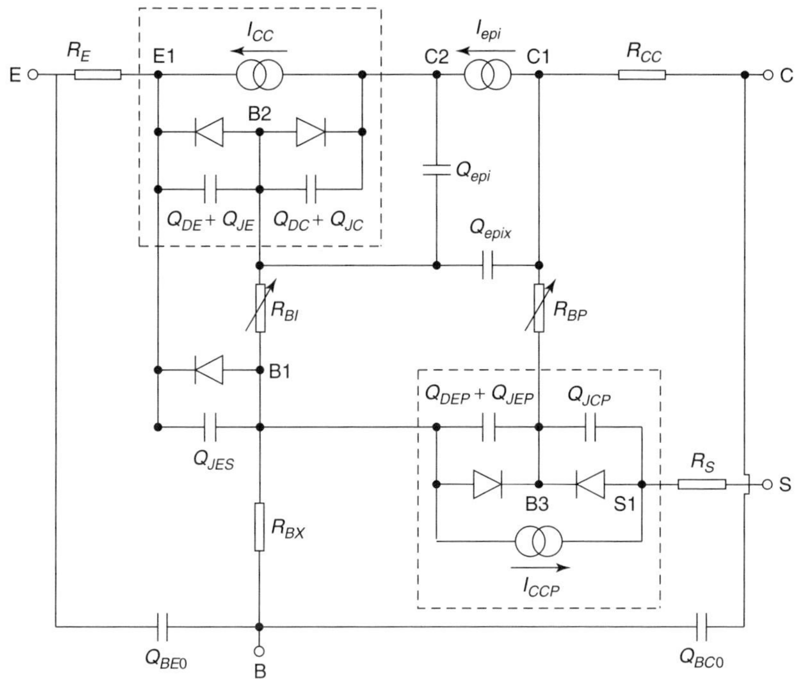

Figure 11.15 Equivalent circuit for the VBIC model [4]

separately from the planar junction using the depletion capacitor Q_{JES} and its associated diode. Overlap capacitances Q_{BE0} and Q_{BC0} are included to respectively model the overlap of the $n+$ polysilicon emitter over the base and the overlap of the $p+$ polysilicon extrinsic base over the collector.

The most important innovation in the VBIC model is improved modelling of the collector epitaxial layer, which is achieved using an approach based on that of Kull *et al.* [14]. The behaviour of the lightly doped collector epitaxial layer is modelled over a wide range of DC and AC operating conditions, and includes the effects of conductivity modulation, velocity saturation, base widening and excess stored charge in the epitaxial collector. The model requires three additional circuit elements I_{EPI}, Q_{epi} and Q_{epix}, as illustrated in Figure 11.15. The current source I_{EPI} models the current in the epitaxial collector region and is given by:

$$I_{epi} = \frac{I_{epi0}}{\sqrt{1 + \left(\frac{I_{epi0}}{I_{epis}}\right)^2}} \tag{11.52}$$

where I_{epi0} represents a model of the epitaxial collector region in the absence of hot carrier effects. The equations for the Kull [14], the VBIC [4] and the Mextram models [5] all reduce to the same equation for I_{epi0} when hot carrier effects are neglected:

$$I_{epi0} = \frac{V_{C1C2} + V_t\left(K_{bci} - K_{bcx} - \ln\left(\frac{K_{bci}+1}{K_{bcx}+1}\right)\right)}{R_{CI}} \tag{11.53}$$

$$k_{bci} = \sqrt{1 + \gamma \exp\frac{V_{B2C2}}{V_t}} \tag{11.54}$$

$$k_{bcx} = \sqrt{1 + \gamma \exp\frac{V_{B2C1}}{V_t}} \tag{11.55}$$

I_{epis} is given by:

$$I_{epis} = \frac{V_0}{R_{CI}} + \frac{0.5\sqrt{V_{C1C2}^2 + 0.01}}{R_{CI}H_{RCF}} \tag{11.56}$$

The VBIC parameters in equations (11.52)–(11.56) are γ ($GAMM$), R_{CI}, H_{RCF} and V_0.

The capacitors labelled Q_{epi} and Q_{epix} model excess stored charge in the epitaxial collector and are given by:

$$Q_{epi} = Q_{C0}K_{bci} \tag{11.57}$$

$$Q_{epix} = Q_{C0}K_{bcx} \tag{11.58}$$

where Q_{C0} is a further VBIC parameter.

11.11 MEXTRAM MODEL

The Mextram bipolar transistor model is a comprehensive, physics-based model that includes all of the important mechanisms in the VBIC model, though the implementation is considerably different. The physical basis of the model means that once the correct parameter set has been found, the description of the device behaviour is sufficiently realistic to allow accurate prediction. The disadvantage is that coupling exists between current and charge models, and hence phenomena like gain roll-off and f_T roll-off cannot be treated separately. This makes parameter extraction somewhat more complex. The Mextram model does not reduce to the standard Gummel-Poon model when parameters are omitted.

A simplified equivalent circuit of the Mextram model is illustrated in Figure 11.16. The behaviour of the lightly doped collector epitaxial layer is modelled using the circuit elements I_{C1C2}, and Q_{epi}, which are described using a similar set of equations to those used in the VBIC model. The Mextram epitaxial layer equations are identical to the VBIC equations when hot carrier effects are neglected, but the treatment of hot carrier effects in the collector epitaxial layer is different [5]. Effects such as conductivity modulation, base widening, velocity saturation and excess stored charge in the epitaxial collector are all accounted for. Substrate effects are modelled using the current source I_{SUB}, and hence the behaviour of the substrate can be modelled using fewer parameters in Mextram than in VBIC. A good comparison of the Mextram and VBIC models is given in [15].

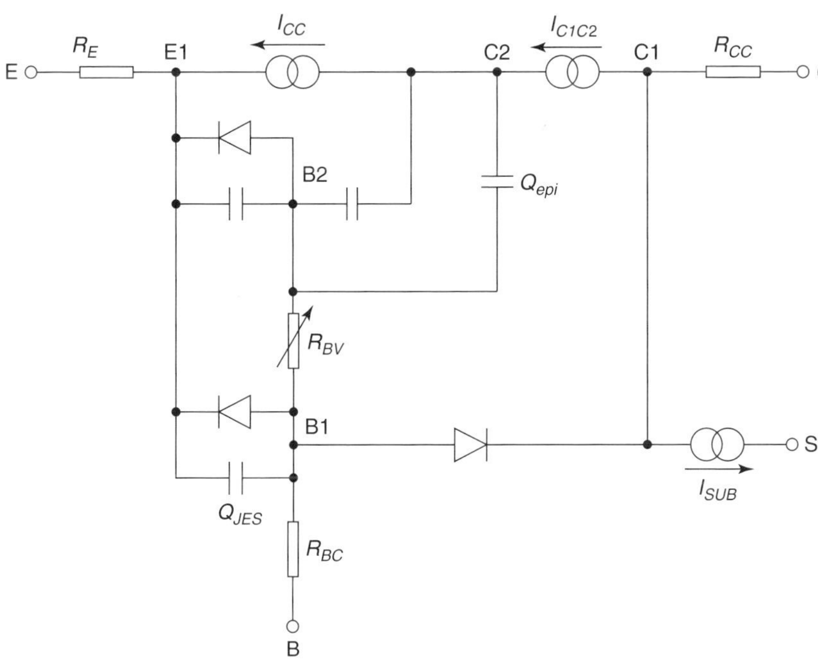

Figure 11.16 Simplified equivalent circuit for the Mextram model [5]

REFERENCES

[1] L.W. Nagel and D.O. Pederson, 'Simulation program with integrated circuit emphasis', *16th Midwest Symposium on Circuit Theory,* 12 April (1973).
[2] J.J. Ebers and J.L. Moll, 'Large signal behaviour of junction transistors', *Proc. IRE*, **42**, 1761 (1954).
[3] H.K. Gummel and H.C. Poon, 'An integral charge control model of bipolar transistors', *Bell Syst. Tech. Jnl*, **49**, 827 (1970).
[4] C.C. McAndrew, J.A. Seitchik, D.F. Bowers, M. Dunn, M. Foisy, E. Getreu, M. McSwain, S. Moinian, J. Parker, D.J. Roulston, M. Schroter, P. van Wijnen and L.F. Wagner, 'VBIC95: the vertical bipolar inter company model', *IEEE Jnl Solid State Circuits*, **31**, 1476 (1995).
[5] H.C. deGraaff and W.J. Kloosterman, 'Modelling of the collector epilayer of a bipolar transistor in the MEXTRAM model', *IEEE Trans. Electron. Devices*, **42**, 274 (1995).
[6] I.E. Getreu, *Modelling the Bipolar Transistor*, Elsevier, Amsterdam (1978).
[7] W.M.C. Sansen and R.G. Meyer, 'Characterisation and measurement of base and emitter resistances of bipolar transistors', *IEEE Jnl Solid State Circuits*, 7, 492 (1972).
[8] P. Spiegel, 'Transistor base resistance and its effect on high speed switching', *Solid State Design*, December 15 (1965).
[9] S.T. Hsu, 'Noise in high gain transistors and its application to the measurement of certain transistor parameters', *IEEE Trans. Electron. Devices*, **18**, 425 (1971).
[10] B. Kulke and S.L. Miller, 'Accurate measurement of emitter and collector series resistance in transistors', *Proc. IRE*, **45**, 90 (1957).
[11] T.H. Ning and D.D. Tang, 'Method for determining the emitter and base series resistance of bipolar transistors', *IEEE Trans. Electron. Devices*, **31**, 409 (1984).
[12] www.semiconductors.philips.com/philips_models
[13] www.designers-guide.com/VBIC/
[14] G.M. Kull, L.W. Nagel, S. Lee, P. Lloyd, E.J. Prendergast and H. Dirks, 'A unified circuit model for bipolar transistors including quasi saturation effects', *IEEE Trans. Electron. Devices*, **32**, 1103 (1985).
[15] L.C.N. deVreede, 'HF silicon IC's for wide-band communication systems', PhD thesis, Technical University of Delft (1996).

12

Optimization of Silicon and Silicon-Germanium Bipolar Technologies

12.1 INTRODUCTION

Optimization of high-speed bipolar processes is a complex task involving interactions between device, process and circuit design. The first problem is identifying an electrical parameter that can serve as a figure of merit for the process performance. Ideally, this parameter should be applicable to all types of circuit, both analogue and digital, and should also be easily measured. Unfortunately, there is no single figure of merit that satisfies these requirements, and in practice a variety of parameters have been used.

The cut-off frequency f_T of a bipolar transistor is often quoted as a figure of merit for a bipolar process. As discussed in Section 5.3, this is related to the forward transit time, τ_F, which is a measure of the fundamental switching speed of the bipolar transistor itself. Unfortunately, the cut-off frequency takes no account of parasitic resistances and capacitances associated with the transistor, and hence is a poor indicator of circuit performance. In order to overcome this limitation the maximum oscillation frequency f_{max} is often quoted as a figure of merit for technology performance. This has been described in Section 5.4 and includes the effects of base resistance and collector capacitance. This is a more realistic figure of merit for analogue circuit performance, although

SiGe Heterojunction Bipolar Transistors Peter Ashburn
 ISBN: 0-470-84838-3

it is questionable whether it is completely accurate for integrated circuit bipolar transistors [1].

The propagation delay, or gate delay, of a logic gate is by far the most reliable figure of merit for digital circuit performance. This is measured on ring oscillators that consist of an odd number of inverters connected together in a ring. On application of power, the circuit begins to oscillate at a frequency equal to twice the propagation delay. Care needs to be taken when interpreting ring oscillator results because ring oscillators are generally designed with minimum load capacitance and unfortunately load capacitance has a big effect on the performance of digital circuits. In particular, load capacitance and fan-out lead to a degradation of the gate delay of bipolar circuits by a factor of approximately 1.5 [2]. In MOS circuits an even greater degradation is obtained by a factor of about 4.5 [2]. However, with the effect of load capacitance properly accounted for, gate delay is an accurate and reliable figure of merit for logic gate performance.

Having selected the propagation delay as an appropriate figure of merit, we next need to devise some means of relating this to the processing parameters. Ideally, we would like to have available an analytical expression for the propagation delay in terms of the processing parameters, such as base implantation dose, emitter/base junction depth, etc. This can be done by defining a quasi-analytical expression for the gate delay in terms of all the time constants of the circuit [3,4]. These time constants are then expressed in terms of transistor parameters, which can be related to the processing parameters through the equations in Chapter 5 or through process and device simulations. This approach provides an effective means of optimizing the process for speed, since the dominant components of the gate delay can be easily identified.

In this chapter we will describe quasi-analytical expressions for the propagation delays of ECL and CML logic gates [3,4]. These expressions will then be used in case studies to optimize the propagation delays of silicon bipolar and SiGe heterojunction bipolar processes [5].

12.2 ECL AND CML PROPAGATION DELAY EXPRESSIONS

Research has shown [6,7] that the propagation delay of bipolar logic circuits can be expressed as a linear combination of the time constants of the circuit, with each time constant weighted by a factor that is dependent upon the circuit configuration. In general, the propagation delay can therefore be approximated by:

$$\tau_D = \sum_i K_i R_i C_i + \sum_j \tau_j \tag{12.1}$$

where K_i and K_j are the weighting factors for the corresponding product terms. The R_i and C_i terms include the resistances and capacitances of the particular logic circuit and the parasitic resistances and capacitances of the bipolar transistor. The τ_j terms are the forward and reverse transit times of the transistor.

Figure 12.1 shows the circuit diagrams for ECL and CML logic gates. The gates can operate from a positive or negative supply voltage. For a positive supply voltage, V_{EE} would be ground and for a negative supply voltage V_{CC} would be ground. The load capacitor C_L is shown for the case of a positive supply voltage. The CML gate comprises a long tail pair and the ECL gate contains in addition an emitter follower. The emitter follower provides a level shift of V_{BE} and ensures that the collector/base junction of the input transistor is never forward biased and hence that it never goes into saturation. The emitter follower also provides strong current drive to the following gates. In the CML gate, the collector/base junction of the input transistor does become forward biased. However, for a typical logic swing of 0.4 or 0.2 V, the forward bias is small and hence the input transistor only goes into weak saturation. The reference voltage V_{REF} is chosen to be halfway between the high and low logic voltages. Both ECL and CML can be considered to be non-saturating

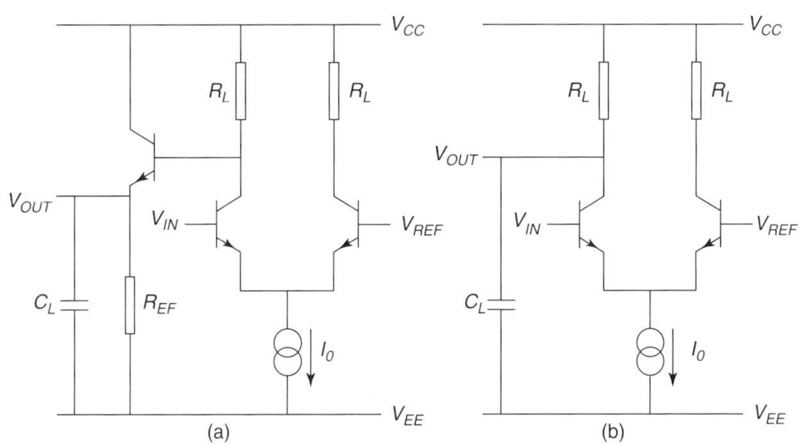

Figure 12.1 Circuit diagrams for (a) an ECL gate and (b) a CML gate

logic gates, and hence their switching speeds are extremely fast. ECL gates have been produced with gate delays of 4.2 ps in a SiGe BiCMOS technology [20].

The gate delays of ECL and CML circuits can be described by the following equation:

$$\begin{aligned}\tau_D = {} & K_1\tau_F + R_L(K_2C_{JCI} + K_3C_{JCX} + K_4C_{JE} + K_5C_{JS} + K_6C_L) \\ & + R_B(K_7C_{JCI} + K_8C_{JCX} + K_9C_{JE} + K_{10}C_{DE} + K_{11}C_{JS} + K_{12}C_L) \\ & + R_C(K_{13}C_{JCI} + K_{14}C_{JCX} + K_{15}C_{JE} + K_{16}C_{DE} + K_{17}C_{JS} + K_{18}C_L) \\ & + R_E(K_{19}C_{JCI} + K_{20}C_{JCX} + K_{21}C_{JE} + K_{22}C_{DE} + K_{23}C_{JS} + K_{24}C_L)\end{aligned} \tag{12.2}$$

where R_L is the load resistance in the ECL or CML circuit, R_{BI} and R_{BX} the intrinsic and extrinsic base resistances, and R_E and R_C the emitter and collector resistances. Similarly, C_{JCI} and C_{JCX} are the intrinsic and extrinsic collector/base depletion capacitances, C_{JE} the emitter/base depletion capacitance, C_{DE} the emitter diffusion capacitance, C_{JCS} the substrate capacitance and C_L the load capacitance due to interconnections and the input capacitance of the following gates. τ_F is the forward transit time. Each of the terms has a weighting factor K associated with it, which can be calculated by applying a sensitivity analysis to SPICE simulation results [4]. The values of the weighting factors are summarized in Table 12.1.

As might be expected, Table 12.1 shows that the important time constants such as R_BC_{JCI} have large weighting factors, whereas less critical time constants such as R_CC_{JE} have much smaller values. Equation (12.2) and the associated time constants are valid for a wide range of parameter values and for a wide range of technologies, including silicon bipolar, SiGe heterojunction bipolar and III/V heterojunction bipolar technologies. Agreement with SPICE simulations is typically better than ±5%.

12.3 CALCULATION OF ELECTRICAL PARAMETERS

The parameter values for insertion in equation (12.1) can be calculated using the equations in Chapter 5, from process and device simulations, or from measured data. The intrinsic and extrinsic base resistances R_{BI} and R_{BX} have been considered in Section 5.6.1, and can be calculated using equations (5.28) and (5.27). The emitter resistance of a bipolar transistor is generally small and dominated by contact resistance, though in polysilicon emitter transistors an additional component is present due

Table 12.1 Summary of the weighting factors for the ECL and CML delay expressions (reprinted with permission from [4])

K	Term	ECL	CML
K_1	τ_F	1.20	1.16
K_2	$R_L C_{JCI}$	1.42	2.52
K_3	$R_L C_{JCX}$	1.50	2.55
K_4	$R_L C_{JE}$	0.04	0.33
K_5	$R_L C_{JS}$	0.27	0.70
K_6	$R_L C_L$	0.31	0.65
K_7	$R_B C_{JCI}$	3.20	3.42
K_8	$R_B C_{JCX}$	0.22	0.26
K_9	$R_B C_{JE}$	0.95	1.01
K_{10}	$R_B C_{DE}$	1.15	1.30
K_{11}	$R_B C_{JS}$	0.14	0.25
K_{12}	$R_B C_L$	0	0.40
K_{13}	$R_C C_{JCI}$	1.72	2.71
K_{14}	$R_C C_{JCX}$	1.62	2.86
K_{15}	$R_C C_{JE}$	0	0.15
K_{16}	$R_C C_{DE}$	0.22	0.30
K_{17}	$R_C C_{JS}$	0.87	0.92
K_{18}	$R_C C_L$	0	0.27
K_{19}	$R_E C_{JCI}$	3.85	3.89
K_{20}	$R_E C_{JCX}$	3.74	3.74
K_{21}	$R_E C_{JE}$	0.38	0.35
K_{22}	$R_E C_{DE}$	0.16	0.79
K_{23}	$R_E C_{JS}$	1.80	0.72
K_{24}	$R_E C_L$	0	0.84

to the resistance of the polysilicon/silicon interface [9]. The collector resistance R_C has been discussed in Section 5.6.2 and can be calculated using equation (5.29). In a well designed bipolar process the collector resistance is determined primarily by the series resistance of the epitaxial collector layer.

The load resistance R_L is determined by the operating conditions of the circuit, and in particular by the logic swing ΔV. In ECL and CML circuits the logic swing is related to the load resistance and collector current through the following equation:

$$R_L = \frac{\Delta V}{I_C} \tag{12.3}$$

The collector current is generally chosen to coincide with the current at which the f_T of the transistor is a maximum, since this maximizes the current that the transistor can deliver to a load. This information can be obtained from a device simulation or from measured data for the transistor of interest.

The capacitances C_{JCI}, C_{JCX}, C_{JE} and C_{JS} are zero bias capacitances and can be calculated using equations analogous to equation (5.30). The split between intrinsic and extrinsic collector capacitance can be estimated from the relative areas of these two regions, as illustrated in Figure 5.9. Emitter diffusion capacitance C_{DE} is a strong function of the collector current, and can be approximated by [4]:

$$C_{DE} = \frac{2I_C\tau_F}{\Delta V} \tag{12.4}$$

The load capacitance C_L is often dominated by the capacitance of the interconnections, which can easily be calculated from the field oxide thickness and the area of the metal tracks. In principle, the forward transit time τ_F can be calculated from equations (5.1), (5.5), (5.8) and (5.11). However, a more accurate value can be obtained from device simulation, and this is advisable since the forward transit time is generally an important component of the gate delay.

12.4 GATE DELAY ESTIMATION

In this section, the gate delay expression in equation (12.2) is first used to calculate an initial value of gate delay of a double polysilicon, self-aligned silicon bipolar transistor, following the approach outlined in [11]. The transistor design was based on that in [10] and used a self-aligned layout, with the critical dimensions given in Table 12.2. Doping concentrations and basewidth were also taken from [10] and are again summarized in Table 12.2.

The transistor electrical parameters were calculated using the approach described in Section 12.3, assuming uniform doping and abrupt junctions for ease of calculation. In calculating the extrinsic base resistance, it was assumed that the extrinsic base polysilicon was silicided and hence the extrinsic base resistance was dominated by intrinsic base region beneath the emitter spacer. The bandgap narrowing model and minority carrier mobility model described in Chapter 3 [12,13] were used to calculate the gain. Device simulations were performed [11] to predict the variation of f_T with collector current and a peak f_T of 15 GHz was obtained at a collector current density of approximately 2×10^4 A/cm^2.

Table 12.2 Summary of dimensions and doping concentrations used in the silicon bipolar transistor design (reprinted with permission from [5])

Parameter	Value
Emitter stripe width	1 μm
Emitter stripe length	5 μm
Emitter spacer thickness	0.2 μm
Extrinsic base stripe length	5 μm
Intrinsic base/collector area	5 μm^2
Extrinsic base/collector area	21 μm^2
Collector/substrate area	95 μm^2
Emitter doping	1×10^{20} cm^{-3}
Base doping	1×10^{18} cm^{-3}
Basewidth	0.10 μm
Collector doping	6×10^{16} cm^{-3}

Table 12.3 Comparison of calculated and measured [10] values of transistor parameters and gate delay for a double polysilicon, self-aligned silicon bipolar transistor (reprinted with permission from [5])

Parameter	Calculated value	Measured value
β	98	100
R_B	220 Ω	150 Ω
R_C	91 Ω	75 Ω
R_E	8.3 Ω	10 Ω
C_{JE}	11.7 fF	19 fF
C_{JC}	19.9 fF	17 fF
C_{JS}	29.3 fF	30 fF
f_T	15 GHz	17–22 GHz
τ_D	42.5 ps	<40 ps

A logic swing of 0.4 V was assumed to calculate the load resistance R_L and a load capacitance of 20 fF was assumed.

Table 12.3 compares the calculated values of transistor electrical parameter with measured values taken from [10]. It can be seen that the calculated values of transistor parameter are in reasonable agreement with the measured values, and in all cases the agreement is better than a factor of two. The calculated value of ECL gate delay τ_D, obtained using equation (12.2), is also in excellent agreement with the measured value reported in [10]. This demonstrates that the gate delay expression gives a reasonably accurate description of the performance of ECL circuits.

12.5 OPTIMIZATION PROCEDURE

To optimize the design of the double polysilicon self-aligned silicon bipolar technology, it is necessary to apply some constraints to ensure that the optimization is realistic. The constraints applied are as follows:

1. minimum gain is 50;
2. maximum doping on low-doped side of junction is $5 \times 10^{18}\,\mathrm{cm}^{-3}$;
3. maximum doping concentration is $1 \times 10^{20}\,\mathrm{cm}^{-3}$;
4. minimum basewidth is determined by base punch-through, calculated with $V_{BE} = 0\,\mathrm{V}$ and $V_{BC} = -5\,\mathrm{V}$;
5. extrinsic base spacer thickness is fixed at $0.2\,\mu\mathrm{m}$;
6. transistor dimensions can be scaled in the same proportion as the emitter width.

The minimum gain of 50 was assumed for correct circuit operation, and the constraint on the doping on the low-doped side of the junction was applied to avoid emitter/base tunnelling leakage due to the creation of a $p+n+$ junction [14].

The procedure used to optimize the gate delay, is to calculate and compare the individual RC time constants to identify the dominant time constant. When this has been done, the transistor can be redesigned to reduce the dominant time constant and hence reduce the gate delay. Recalculation of the individual time constants will then lead to a new dominant time constant, which can in turn be reduced by re-designing the transistor. Full optimization of the transistor design can be achieved by further iterations around this loop.

12.6 OPTIMIZATION OF SILICON BIPOLAR TECHNOLOGY

The trade-offs in the design of the doping profiles of a silicon bipolar transistor are shown in Figure 12.2, where the gain is plotted against base doping for a range of basewidths. For the widest basewidths, the $\beta > 50$ criterion determines the highest value of base doping that can be used, which is around $8 \times 10^{17}\,\mathrm{cm}^{-3}$ for a basewidth of $0.2\,\mu\mathrm{m}$ and $3 \times 10^{18}\,\mathrm{cm}^{-3}$ for a basewidth of $0.1\,\mu\mathrm{m}$. For the thinnest basewidths, the gain is increased sufficiently that the maximum base doping is determined by the onset of tunnelling leakage, which limits the base doping to around

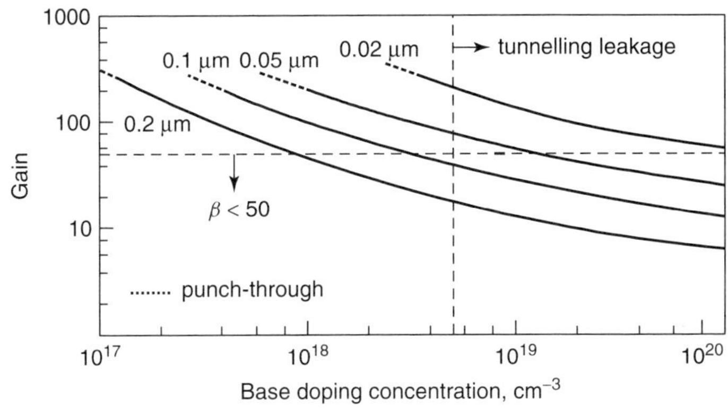

Figure 12.2 Gain of a silicon bipolar transistor as a function of base doping for various basewidths. The emitter doping is fixed at 1×10^{20} cm^{-3} and the emitter depth at 0.2 μm (reprinted with permission from [5])

5×10^{18} cm^{-3}. For a basewidth of 0.02 μm, the base doping at which punch-through occurs almost coincides with the base doping at which tunnelling leakage becomes a problem. Hence basewidths at or below 0.02 μm are not practical for silicon bipolar transistors.

Figure 12.3 shows the variation in propagation delay with basewidth for various values of base doping. The emitter doping and emitter depth

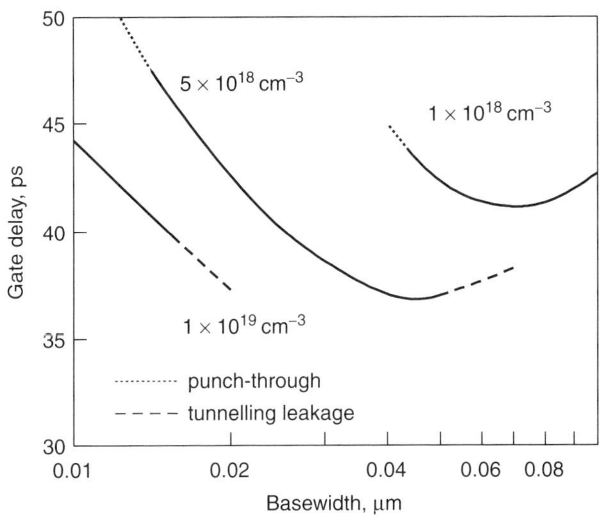

Figure 12.3 Gate delay of an ECL circuit implemented in a double polysilicon, self-aligned silicon bipolar technology as a function of basewidth for various base doping concentrations (reprinted with permission from [5])

were chosen to give an optimum propagation delay at each point. Tunnelling leakage imposes a constraint on the maximum base doping that can be achieved. Note that the base doping of $1 \times 10^{19}\,cm^{-3}$ is possible at small basewidths because the narrow basewidth gives a high enough gain to enable the emitter doping to be reduced below $5 \times 10^{18}\,cm^{-3}$, thereby avoiding emitter/base tunnelling leakage. A base doping of $1 \times 10^{20}\,cm^{-3}$ is not possible for any of the basewidths considered. Punch-through limits the basewidth at small values of basewidth. The minimum gate delay is 37 ps at a basewidth of 0.05 μm, an emitter depth of 0.15 μm, an emitter doping of $1 \times 10^{20}\,cm^{-3}$ and a base doping of $5 \times 10^{18}\,cm^{-3}$.

To gain insight into the limiting time constants, Table 12.4 gives a breakdown of the time constants for a basewidth of 0.1 μm. The total delay is the sum of the individual time constant terms and a contribution of 8.4 ps from the forward transit time. At the end of each row and the bottom of each column, the contribution from each electrical parameter is shown. The table shows that the limiting parameter is the load resistance R_L which comprises 42% of the total gate delay. The second largest contribution to the gate delay comes from the extrinsic collector/base capacitance, C_{JCX}. The extrinsic collector/base capacitance terms comprise 31% of the total gate delay. It is clear from Table 12.4 that a significant reduction of gate delay could be achieved if the load resistance and extrinsic collector/base capacitance could be reduced.

There are three possible ways of reducing the load resistance. The first would be to simply increase the emitter stripe length, hence increasing the emitter area and increasing the collector current in equation (12.3). Unfortunately, this approach increases the parasitic depletion capacitances, and hence negates the benefits of the reduced load resistance. The second approach is to reduce the logic swing of the ECL gate. The minimum value of logic swing is determined by the noise immunity

Table 12.4 Breakdown of gate delay time constants for an ECL circuit implemented in a double polysilicon, self-aligned silicon bipolar technology. The transistor has a basewidth of 0.1 μm. The total gate delay includes a contribution of 8.4 ps from the forward transit time (reprinted with permission from [19])

	C_{JCI}	C_{JCX}	C_{JE}	C_{DE}	C_{JS}	C_L	Total, ps
R_B	1.41	0.42	2.00	4.60	0.47	–	8.9
R_L	2.18	9.86	0.29	–	3.16	2.48	17.97
R_C	0.60	2.43	–	0.70	2.32	–	6.05
R_E	0.12	0.51	0.06	0.05	0.44	–	1.18
Total, ps	4.31	13.22	2.35	5.35	6.39	2.48	42.5

requirements of the circuit. A value of 0.4 V was used in the calculation, though values as low as 0.2 V have been used in practice [15]. The third approach is to design transistors that operate at higher values of collector current. As discussed in Section 5.5, the Kirk effect limits the collector current at which the transistor is capable of operating. Higher values of collector current can be achieved by increasing the collector doping concentration, as shown in equation (5.26), to push the Kirk effect to higher currents.

Figure 12.4 shows the variation of gate delay with collector current density for two values of logic swing. For a logic swing of 0.4 V, the gate delay decreases as the collector current density is increased above its previous value of 2×10^4 A/cm^2, reaching at minimum of 29.2 ps at a current density of around 6×10^4 A/cm^2. This decrease is due to the decrease in the load resistance terms in Table 12.4. At higher collector current densities, the gate delay increases. This behaviour can be explained by the diffusion capacitance (C_{DE}) terms, which increase with increasing collector current, as can be seen from equation (12.4). Similar behaviour is seen for a logic swing of 0.2 V, but the minimum value of gate delay occurs at a lower value of collector current density. This behaviour can be explained by noting that the load resistance

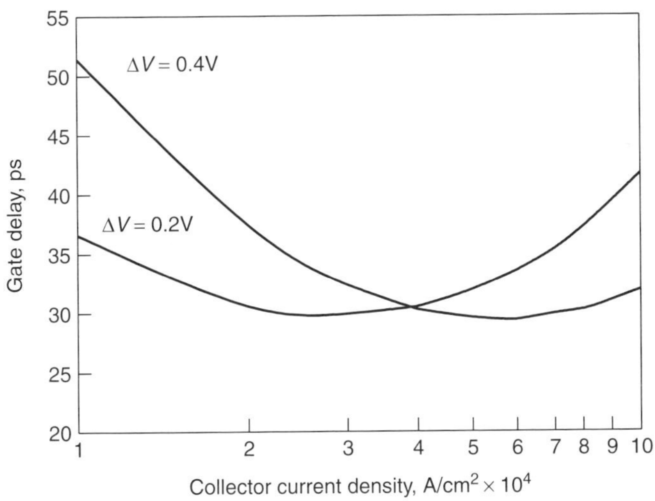

Figure 12.4 Gate delay as a function of collector current density for an ECL circuit implemented in a double polysilicon, self-aligned silicon bipolar technology, for logic swings of 0.4 and 0.2 V. The transistor has an emitter doping of 1×10^{20} cm^{-3}, a base doping of 5×10^{18} cm^{-3}, and emitter depth of 0.15 μm and a basewidth of 0.05 μm (reprinted with permission from [5])

required for a given collector current density using a 0.2 V logic swing is half that required for a 0.4 V logic swing.

The second largest contribution to the total gate delay in Table 12.4 comes from the extrinsic collector/base capacitance C_{JCX}, which comprises 31% of the total gate delay. One obvious way of reducing the extrinsic base capacitance, and at the same time the other depletion capacitances, is to scale the transistor geometry to smaller dimensions. The effect of scaling is illustrated in Figure 12.5 for a logic swing of 0.4 V. A pronounced improvement in gate delay is obtained on scaling the dimensions from 1 μm to 0.2 μm, with the minimum delay being 14.2 ps at a collector current density of 10^5 A/cm^2. This gate delay does not represent the minimum achievable in a double polysilicon self-aligned silicon bipolar technology, since further improvements could be obtained by scaling the transistors in a way that better addressed the dominant time constants, particularly those involving the extrinsic collector/base capacitance. The extrinsic base area, and hence the extrinsic collector/base capacitance, is determined by the overlap of the $p+$ polysilicon extrinsic base over the active area and by the emitter spacer thickness, as illustrated in Figure 9.10(a). Reduction of these layout parameters would deliver a further reduction in gate delay. In practice gate delays

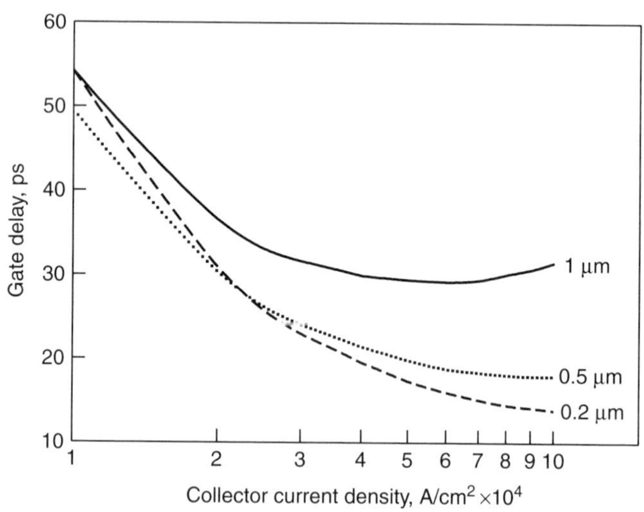

Figure 12.5 Gate delay as a function of collector current density for an ECL circuit implemented in a double polysilicon, self-aligned silicon bipolar technology, for a logic swing of 0.4 V and for three different transistor geometries. The transistor has an emitter doping of 1×10^{20} cm^{-3}, a base doping of 5×10^{18} cm^{-3}, and emitter depth of 0.15 μm and a basewidth of 0.05 μm (reprinted with permission from [5])

down to 12 ps [16] have been reported for ECL circuits implemented in a double polysilicon, self-aligned silicon bipolar technology.

12.7 OPTIMIZATION OF SILICON-GERMANIUM HBT TECHNOLOGY

The transistor electrical parameters were calculated using the same transistor layout and using the method described in Section 12.5. The doping and germanium profiles were assumed to be uniform and the junctions abrupt for ease of calculation. The doping-induced bandgap narrowing and minority carrier mobility in the SiGe base were assumed to be the same as for a Si base, and the models described in Chapter 3 [12,13] were used. The germanium-induced bandgap narrowing in the base was calculated using the values in [17] and [18]. A logic swing of 0.4 V was assumed and a load capacitance of 20 fF.

The trade-offs in the design of the doping profiles of a SiGe HBT are illustrated in Figure 12.6, which shows a graph of current gain against base doping for various values of basewidth and for a germanium content of 12%. The main benefit of a SiGe HBT is that the high gain resulting from the inclusion of germanium in the base can be traded for increased base doping concentration. Furthermore the emitter doping concentration can be reduced to avoid emitter/base tunnelling leakage. In Figure 12.6, an emitter doping concentration of $1 \times 10^{18}\,cm^{-3}$ has been used, and as a result the base doping can be increased to $1 \times 10^{20}\,cm^{-3}$

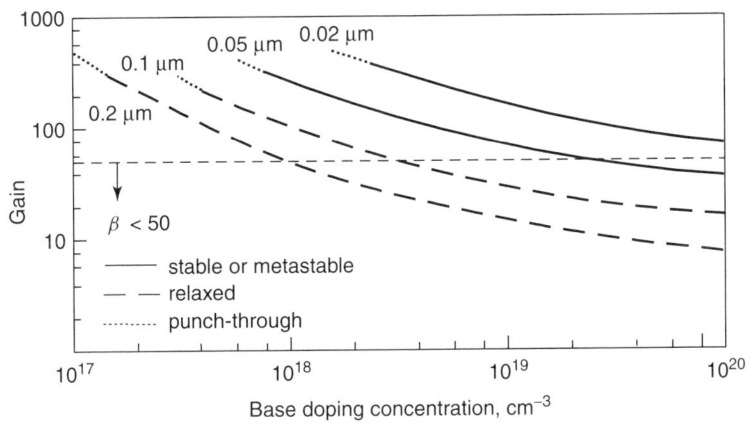

Figure 12.6 Gain of a SiGe HBT as a function of base doping for various basewidths. The emitter doping was fixed at $1 \times 10^{18}\,cm^{-3}$, the emitter depth at 0.2 μm, and the germanium content at 12% (reprinted with permission from [5])

without encountering any problems with emitter/base tunnelling leakage. For values of basewidth ≥ 0.05 μm, the constraint on the minimum gain limits the maximum base doping. However for the thinnest basewidth of 0.02 μm, the base doping can be increased to above 1×10^{20} cm^{-3} without violating the minimum gain criterion. This increased degree of freedom in the choice of the base doping and basewidth creates additional opportunities for increasing the f_T and reducing the gate delay. Punch-through of the base limits the minimum base doping and also the minimum basewidth, though even for the smallest basewidth of 0.02 μm, a wide range of base doping concentrations can be chosen. Basewidths below 0.01 μm can be used if the base doping concentration is high. Finally, the two largest basewidths of 0.2 and 0.1 μm are well above the critical thickness for a germanium content of 12%, and hence would be relaxed and unusable. The 0.05 μm base is just above the metastable line and the 0.02 μm base is stable, as can be seen by referring to Figure 7.3.

Figure 12.7 shows a graph of the propagation delay as a function of basewidth for various values of base doping. The emitter doping and emitter depth were chosen to give an optimum propagation delay at each

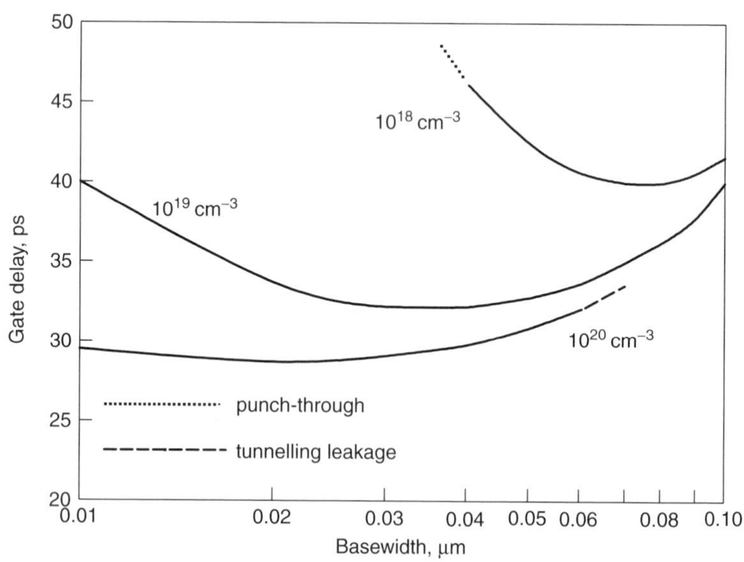

Figure 12.7 Gate delay of an ECL circuit implemented in a double polysilicon, self-aligned SiGe HBT technology as a function of basewidth for various base doping concentrations. The germanium content was 12%, and the emitter doping and emitter depth were chosen to give an optimum gate delay at each point (reprinted with permission from [5])

point. For a low base doping concentration of $1 \times 10^{18}\,\text{cm}^{-3}$, punch-through limits the extent that the basewidth can be reduced, whereas for base doping concentrations of $1 \times 10^{19}\,\text{cm}^{-3}$ and $1 \times 10^{20}\,\text{cm}^{-3}$, punch-through is not a problem even at a basewidth of 0.01 μm. The gain criterion becomes difficult to meet at the highest base doping concentration of $1 \times 10^{20}\,\text{cm}^{-3}$ and at the widest basewidths. For these conditions, tunnelling leakage is significant, since it is necessary to increase the emitter doping to achieve a gain of greater than 50. A minimum gate delay is achieved for a given base doping, which represents a compromise between high base resistance at narrow basewidths and high forward transit time at wider basewidths. The minimum gate delay of 28.5 ps is achieved at a basewidth of 0.02 μm and a base doping concentration of $1 \times 10^{20}\,\text{cm}^{-3}$.

To gain insight into the time constants that limit the circuit performance, Table 12.5 gives a breakdown of the time constants for an ECL circuit incorporating SiGe HBTs with a basewidth of 0.02 μm, an emitter doping of $1 \times 10^{18}\,\text{cm}^{-3}$, a base doping of $1 \times 10^{20}\,\text{cm}^{-3}$ and an emitter depth of 0.2 μm. The total delay is the sum of the individual time constant terms and a contribution of 2.0 ps from the forward transit time. The table shows that the largest contribution to the gate delay comes from the load resistance R_L, and the second largest contribution from the extrinsic collector/base capacitance C_{JCX}. This situation is very similar to that found for the silicon bipolar transistor in Table 12.4, and hence it is clear that a significant reduction of gate delay could be achieved if the load resistance and extrinsic collector/base capacitance could be reduced.

As shown by equation (12.3), the load resistance can be decreased by increasing the collector current at which the transistor is capable of

Table 12.5 Breakdown of gate delay time constants for an ECL circuit implemented in a double polysilicon, self-aligned SiGe HBT technology. The transistor has a germanium content of 12%, basewidth of 0.02 μm, an emitter doping of $1 \times 10^{18}\,\text{cm}^{-3}$, a base doping of $1 \times 10^{20}\,\text{cm}^{-3}$ and an emitter depth of 0.2 μm. The total gate delay includes a contribution of 2.0 ps from the forward transit time (reprinted with permission from [19])

	C_{JCI}	C_{JCX}	C_{JE}	C_{DE}	C_{JS}	C_L	Total, ps
R_B	0.26	0.07	0.33	0.16	0.09	–	0.91
R_L	2.00	9.02	0.24	–	3.16	2.48	16.90
R_C	0.48	1.94	–	0.10	2.03	–	4.55
R_E	0.21	0.86	0.09	0.01	0.81	–	1.98
Total, ps	2.95	11.89	0.66	0.27	6.09	2.48	25.7

operating, which is limited by the Kirk effect, as discussed in Section 5.5. Higher values of collector current can be achieved by increasing the collector doping concentration, as shown in equation (5.26), to push the Kirk effect to higher currents. Figure 12.8 shows the variation of ECL gate delay with collector current density for two values of logic swing. For a logic swing of 0.4 V, the gate delay decreases as the collector current density is increased, reaching at minimum of 12.9 ps at a current density of around 1.6×10^5 A/cm^2. This decrease is due to the decrease in the load resistance terms in Table 12.4. At higher collector current densities, the gate delay increases. This behaviour can be explained by the diffusion capacitance (C_{DE}) terms, which increase with increasing collector current, as can be seen from equation (12.4). Similar behaviour is seen for a logic swing of 0.2 V, but the minimum value of gate delay occurs at a lower value of collector current density.

The extrinsic collector/base capacitance can be reduced by scaling the transistor geometry to smaller dimensions. The effect of scaling is illustrated in Figure 12.9 for a logic swing of 0.4 V. A pronounced improvement in gate delay is obtained on scaling the dimensions from 1 μm to 0.2 μm, with the minimum delay being 7 ps at a collector current density of around 3×10^5 A/cm^2. This gate delay does not represent the minimum achievable in a double polysilicon self-aligned SiGe HBT

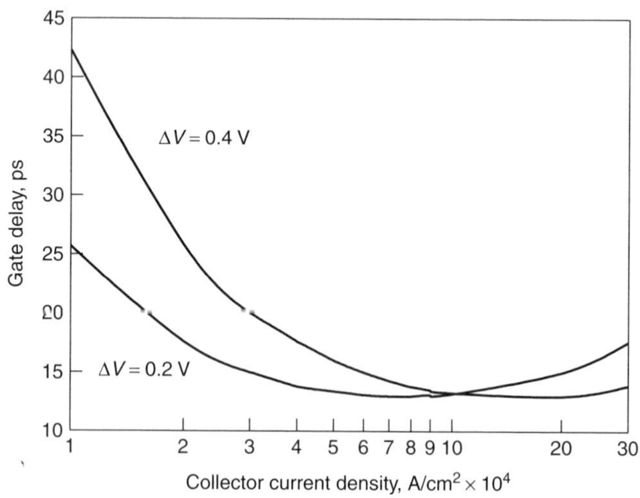

Figure 12.8 Gate delay as a function of collector current density for an ECL circuit implemented in a double polysilicon, self-aligned SiGe HBT technology, for logic swings of 0.4 and 0.2 V. The transistor has a germanium content of 12%, a basewidth of 0.02 μm, an emitter doping of 1×10^{18} cm^{-3}, a base doping of 1×10^{20} cm^{-3} and an emitter depth of 0.2 μm (reprinted with permission from [19])

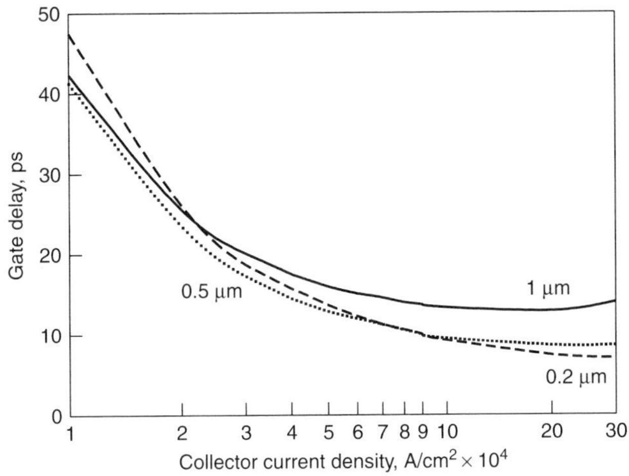

Figure 12.9 Gate delay as a function of collector current density for an ECL circuit implemented in a double polysilicon, self-aligned SiGe HBT technology, for a logic swing of 0.4 V and for three different transistor geometries. The transistor has a germanium content of 12%, a basewidth of 0.02 μm, an emitter doping of $1 \times 10^{18}\,cm^{-3}$, a base doping of $1 \times 10^{20}\,cm^{-3}$ and an emitter depth of 0.2 μm (reprinted with permission from [19])

technology, since constraints have been applied to the optimization and scaling procedures. For example, further improvements could be obtained by grading the germanium profile and by more aggressively scaling the area of the extrinsic base junction and hence of the extrinsic collector/base capacitance. In practice, gate delays down to 4.2 ps [20] have been reported for ECL circuits implemented in a double polysilicon, self-aligned SiGe HBT technology.

REFERENCES

[1] G.W. Taylor and J.G. Simmons, 'Figure merit for integrated bipolar transistors', *Solid State Electronics*, **29**, 941 (1986).

[2] A.D. Welbourn, 'Gigabit logic', *IEE Proc., Pt I*, **129**, 157 (1983).

[3] E.F. Chor, A. Brunnschweiler and P. Ashburn, 'A propagation delay expression and its application to the optimization of polysilicon emitter ECL processes', *IEEE Jnl Solid State Circuits*, **23**, 251 (1988).

[4] W. Fang, 'Accurate analytical delay expressions for ECL and CML circuits and their applications to optimising high-speed bipolar circuits', *IEEE Jnl Solid State Circuits*, **25**, 572 (1990).

[5] Z.A. Shafi, P. Ashburn and G.J. Parker, 'Predicted propagation delay of Si/SiGe heterojunction ECL circuits', *IEEE Jnl Solid State Circuits*, **25**, 1268 (1990).

[6] D.D. Tang and P.M. Solomon, 'Bipolar transistor design for optimized power- delay logic circuits', *IEEE Jnl Solid State Circuits*, **14**, 679 (1979).
[7] R. Ranfft and H.M. Rein, 'A simple optimization procedure for bipolar subnanosecond ICs with low power dissipation', *Microelectronics Jnl*, **13**, 23 (1982).
[8] K. Washio, H. Shimamoto, K. Oda, R. Hayami, Y. Kiyota, M. Tanabe, M. Kondo, T. Hashimoto and T. Harade, 'A 0.2 μm 180 GHz f_{max} 6.7 ps ECL SOI/HRS self-aligned SEG SiGe HBT/CMOS technology for microwave and high-speed digital applications', *IEEE Trans. Electron. Devices*, **49**, 271 (2002).
[9] E.F. Chor, P. Ashburn and A. Brunnschweiler, 'Emitter resistance of arsenic and phosphorus doped polysilicon emitter transistors', *IEEE Electron. Device Lett.*, **6**, 516 (1985).
[10] M.C. Wilson, P.C. Hunt, S. Duncan and D.J. Bazley, '10.7 GHz frequency divider using double layer silicon bipolar process technology', *Electron. Lett.*, **24**, 920 (1988).
[11] Z.A. Shafi, P. Ashburn and G.J. Parker, 'Predicted propagation delay of Si/SiGe heterojunction bipolar ECL circuits', *IEEE Journal Solid State Circuits*, **25**, 1268 (1990).
[12] S.E. Swirhun, Y.H. Kwark and R.M. Swanson, 'Measurement of electron lifetime, electron mobility, and bandgap narrowing in heavily doped p-type silicon', *IEDM Technical Digest*, **24** (1986).
[13] J. Del Alamo, S. Swirhun and R.M. Swanson, 'Measuring and modeling minority carrier transport in heavily doped silicon', *Solid State Electronics*, **28**, 47 (1985).
[14] J.A. del Alamo and R.M. Swanson, 'Forward biased tunnelling current: a limitation to bipolar device scaling', *IEEE Electron Device Letters*, **7**, 629 (1986).
[15] B. Hoffmann, H. Klose, T. Meister, I. Kerner and R. Schreiter, 'A 10 GHz high performance BiCMOS technology for mixed signal CMOS/ECL ic's', *Proc. ESSDERC*, 477 (1989).
[16] J. Böck, H. Knapp, K. Aufinger, M. Wurzer, S. Boguth, R. Schreiter, T.F. Meister, M. Rest, M. Ohnemus and L. Treitinger, '12 ps implanted base silicon bipolar technology', *Proc. BCTM* (1999).
[17] G.L. Patton, S.S. Iyer, S.L. Delage, S. Tiwari and J.C. Stork, 'Silicon-germanium base heterojunction bipolar transistors by molecular beam epitaxy', *IEEE Electron. Device Lett.*, **9**, 165 (1988).
[18] C.A. King *et al.*, 'Si/SiGe heterojunction bipolar transistors produced by limited reaction processing', *IEEE Electron. Device Lett.*, **10**, 52 (1989).
[19] Z.A. Shafi, 'An investigation of silicon heterojunction bipolar transistors', PhD thesis, University of Southampton (1992).
[20] J. Jagannathan, M. Meghelli, A.V. Rylyakov, R.A. Groves, A.K. Chinthakindi, C.M. Schnabel, D.A. Ahlgren, G.G. Freeman, K.J. Stein and S. Subbanna; 'A 4.2 ps ECL ring oscillator in a 285 GHz f_{max} SiGe technology', *IEEE Electron. Device Lett.*, **23**, 541 (2002).

Index

SiGe Heterojunction Bipolar Transistors Peter Ashburn

ISBN: 0-470-84838-3